# HOT CONNECTIONS

# HOT CONNECTIONS

### Aluminum Wire,
### Beverly Hills Supper Club Fire, and
### the Myth of Self-Regulating Industry

Jesse Aronstein

LONDONDERRY PRESS ❦ SCHENECTADY, NEW YORK

Copyright ©2021 Jesse Aronstein

All rights reserved under International and Pan-American Copyright Conventions. No part of this book may be used or reproduced in any manner whatsoever without written permission from the publisher, except in the case of brief quotations used in critical articles and reviews.

ISBN: 978-0-578-89835-3 (print)

Also available in e-book

Cover and Book Design by Sue Campbell

Londonderry Press, Schenectady, New York
AronsteinJ@aluminumwire.info

# CONTENTS

Abbreviations and Acronyms . . . . . . . . . . . . . . . . . . . . . . . . . . . . . . . . . . 6
Introduction . . . . . . . . . . . . . . . . . . . . . . . . . . . . . . . . . . . . . . . . . . . . . . . . . 7
   1. The Beverly Hills Supper Club Fire . . . . . . . . . . . . . . . . . . . . . . . . . . 9
   2. Fatal Fire—Hampton Bays, New York . . . . . . . . . . . . . . . . . . . . . . 23
   3. Kaiser Aluminum and Chemical Corp. . . . . . . . . . . . . . . . . . . . . . 29
   4. Fatal Fire—Columbus, Georgia . . . . . . . . . . . . . . . . . . . . . . . . . . . 41
   5. "Shootout" at Travelers Hotel . . . . . . . . . . . . . . . . . . . . . . . . . . . . . 51
   6. A Fire Chief's Perspective . . . . . . . . . . . . . . . . . . . . . . . . . . . . . . . . 61
   7. Underwriters Laboratories . . . . . . . . . . . . . . . . . . . . . . . . . . . . . . . 67
   8. Mobile Home Fires . . . . . . . . . . . . . . . . . . . . . . . . . . . . . . . . . . . . . 77
   9. Trial by Television . . . . . . . . . . . . . . . . . . . . . . . . . . . . . . . . . . . . . . 91
  10. Fatal Fire—Houston, Texas, 1978 . . . . . . . . . . . . . . . . . . . . . . . . 105
  11. The UL Ad Hoc Committee . . . . . . . . . . . . . . . . . . . . . . . . . . . . . 111
  12. The Consumer Product Safety Commission . . . . . . . . . . . . . . . . 133
  13. "If Properly Installed" . . . . . . . . . . . . . . . . . . . . . . . . . . . . . . . . . . 153
  14. Fatal Fire—Phoenix Arizona . . . . . . . . . . . . . . . . . . . . . . . . . . . . 169
  15. Canada . . . . . . . . . . . . . . . . . . . . . . . . . . . . . . . . . . . . . . . . . . . . . . 175
  16. An Industry on Trial . . . . . . . . . . . . . . . . . . . . . . . . . . . . . . . . . . . 189
  17. Arrested Development . . . . . . . . . . . . . . . . . . . . . . . . . . . . . . . . . 201
  18. The Myth of the Self-Regulating Industry . . . . . . . . . . . . . . . . . . 217
Postscript . . . . . . . . . . . . . . . . . . . . . . . . . . . . . . . . . . . . . . . . . . . . . . . . . 223
Additional Resource for Aluminum Wiring Information . . . . . . . . . . . . . . . . 226
About the Author . . . . . . . . . . . . . . . . . . . . . . . . . . . . . . . . . . . . . . . . . . 227
References, Sources, and Notes . . . . . . . . . . . . . . . . . . . . . . . . . . . . . . . 229
Index . . . . . . . . . . . . . . . . . . . . . . . . . . . . . . . . . . . . . . . . . . . . . . . . . . . . 251

# ABBREVIATIONS and ACRONYMS

## ENTITIES

| | |
|---|---|
| AA | Aluminum Association |
| CEA | Canadian Electrical Association |
| CIA | US Central Intelligence Agency |
| CPSC | Consumer Product Safety Commission |
| CSA | Canadian Standards Association |
| EEI | Edison Electric Institute |
| FCC | US Federal Communications Commission |
| IAEI | International Association of Electrical Inspectors |
| FRI | Franklin Research Institute (later FRC, Franklin Research Center) |
| NBS | National Bureau of Standards (later Nat. Inst. of Standards & Technology) |
| NEMA | National Electrical Manufacturers Association |
| NFPA | National Fire Protection Association |
| PLCC | Plaintiffs' Lead Council Committee (Beverly Hills Supper Club litigation) |
| UL or U.L. | Underwriters Laboratories Inc. |
| USDA | US Dept. of Agriculture |
| W-M | Wright-Malta Corp. |

## OTHER

| | |
|---|---|
| ALUMICON | brand of connector (King Innovations) |
| AL-CU (or AL/CU, CU-AL, CU/AL) | mark signifying UL or CSA approval for use with aluminum and/or copper wire (until 1973) |
| AWG | American Wire Gage |
| CO/ALR | mark signifying UL or CSA approval for use with aluminum and/or copper wire (since 1973) |
| COPALUM | brand of connector (AMP, later Tyco, then TE Connectivity) |
| EC | Electrical Conductor grade of aluminum wire. |
| EEE | brand of aluminum alloy wire (Southwire) |
| KA-FLEX | brand of NM cable with aluminum wire (Kaiser) |
| NEC | National Electrical Code (in US) |
| NM | Non-Metallic sheathed cable (commonly called "Romex") |

# INTRODUCTION

The catastrophic fire at the Beverly Hills Supper Club in Southgate, Kentucky in 1977 resulted in 165 fatalities. About 2,400 people were in the building. People came to hear John Davidson sing, to have dinner and to attend a variety of functions, including meetings and a wedding reception. This fire ranks high among the major civilian fire losses of the 20th century.

Lawsuits followed. An aluminum wire connection failure was alleged to be the cause of the fire. Thirty-six companies in the electrical industry defended their products and their actions against claims that the aluminum wiring system they marketed was inherently dangerous, that they had conspired to market it in spite of their knowledge of the hazard and that they failed to warn the public.

The information that came to light in that litigation tells the story of a self-regulating industry that managed an electrical fire safety problem to protect itself and not the public. Once in motion, the marketing of aluminum wiring was a rolling stone that could not be stopped by well-intended individuals in the companies that were involved. The industry's standards and practices were not equal to the task. Corporate interests and individual careers were protected at the expense of public safety.

Six years before the deadly fire, at about the same time that some aluminum wiring was being installed in the club, R.J. Schoerner, Vice President of one of the aluminum wire manufacturers, wrote:

> *A genuine crisis exists in our industry, of a magnitude that jeopardizes the entire aluminum conductor market and the life and property of consumers.*

This is the story of how that came to be, and how—if at all—it was resolved. Its foundation is largely built on documents that surfaced in the Beverly Hills Supper Club fire litigation. It is appropriate, therefore, to dedicate this book to the memory of those who died in that disaster.

# 1.
# THE BEVERLY HILLS SUPPER CLUB FIRE

The Beverly Hills Supper Club in rural Southgate, Kentucky was situated on the side of a hill at the end of a long, curved driveway. It stood three miles southwest of Cincinnati, across the Ohio River. The Club's main building covered more than one acre. Seen from the driveway, an elegant facade hid the original plain brick building from view. The facade reflected a sort of Las Vegas entertainment style; fancy and inviting, a place apart from the ordinary. Behind the main building was a garden. A chapel, used mainly for wedding ceremonies, stood at one far corner of the garden. At the other far corner was a gazebo.

The large and elegantly decorated building could host a variety of activities simultaneously. On busy evenings in the mid-1970s, the club hosted upscale diners and drinkers, meetings, receptions, dances, and entertainment. Milton Berle, Pat Boone, The Mills Brothers, and Ray Charles were among the well-known entertainers who performed in the Club's Cabaret Room.

On the night of the fire, May 28, 1977, there was a small party in one section of the Viennese Room, a wedding reception in the Zebra Room and a Savings and Loan Association's awards banquet in the Empire Room. In the two large Crystal Rooms upstairs were a dog club's party and a choral group's dinner, style show and dance. Featured in the Cabaret Room was John Davidson, a popular singer. His two performances were scheduled to start at 8:30 and 10:30, each with an audience warm-up by a comedy duo,

Jim Teeter and Jim McDonald. Elsewhere in the club, dinner and drinks were being served.

This was Saturday night of Memorial Day weekend. The Cabaret Room was packed with more than 1,200 patrons for John Davidson's first show. Many had dined elsewhere in the Club and then had gone to the Cabaret Room early to get good seats. Another 1,200 or so patrons and employees were in less crowded dining rooms, party rooms and the working areas of the club.

Earlier that day, a small fire—just a smoldering ember—ignited in a concealed wall or ceiling space adjacent to the Zebra Room. Most employees started work in the afternoon. Patrons started coming at about 5:30. The smoldering fire grew as they arrived, unobserved and unobservable.

One employee in the main bar area recalled smelling a whiff of smoke sometime between 4:00 and 5:00. At the time, he thought that it probably came from the kitchen. A couple waiting in the vehicle line on the long driveway noticed a brief cloud of smoke drift up from the roof shortly after 6:00. Another patron arriving at about 7:00 noticed some smoke coming from the roof, but decided it was nothing to be concerned about.

Sometime after 8:00 that evening, the Zebra Room wedding reception concluded. Some of the guests had complained that the room was very hot. Employees who briefly entered the empty Zebra Room after the reception saw no indication of fire. Above the Zebra Room, the fashion models preparing for the choral group's style show felt very hot in the dressing room.

The smoldering fire had grown slowly. When flame eventually erupted, it spread rapidly. At about 8:50, waitresses Roberta Vanover and her sister Marsha entered the Zebra Room through its main entrance door, saw dense smoke at the ceiling level, and alerted other employees. Eileen Druckman, the telephone reservationist working in the "cubbyhole" between the main bar and the Zebra Room, smelled smoke and opened the door to the Zebra Room. She saw the smoke and promptly alerted a nearby bartender and the Club hostess.

# 1. THE BEVERLY HILLS SUPPER CLUB FIRE

Rick Shilling, one of the owner's sons, quickly went to the Zebra Room, intending to find and suppress the fire with portable extinguishers. He quickly called the fire department when he realized that the staff could not deal with the situation. The Southgate Fire Department logged the first call at 9:01. Three fire departments responded on the first alarm. The first fire and rescue apparatus arrived at the club a few minutes later.

Individual employees and patrons alerted customers and staff, mainly on their own initiative. There was no general fire alarm system. No emergency evacuation plan had been developed. The staff had not been trained to deal with such a situation. Eventually, however, many employees understood that the building itself was on fire and had to be evacuated. A waitress in the Cafe dining room recalled her personal experience. She had just taken a drink order.

> *I had turned around to walk away from the table, and I looked up front, which after you go to a certain point in the dining room, you could see straight up to the bar, and the hostesses stand, and the busboy started by and I said to him, 'Look at all that smoke. The exhaust fans must have gone off.' I thought this must be what had happened. And I thought, 'Well a trash can has probably caught on fire' ... and he looked up and said, 'Yeah.'*
> 
> *... So I started up front to see if there was a garbage (can) on fire, so maybe I could grab a pitcher of water to throw on it. And as I started up front, I noticed the smoke was a little bit thicker, so I thought, 'Boy, they've really got a garbage can going up there.' And about that time, there were customers seated in the front part of that room. See the dining room is like this. In the very middle there is a big fireplace. On this side of the fireplace, there is a service stand where you get your silverware, cups, and saucers and water. You can't see all of these people seated there if you're in behind, so as I started around, people jumped up and started running. I thought,*

*Gee, that's kind of dumb.' So I turned back around because a woman screamed and she jumped up and I said, 'Everybody be seated. There's nothing to worry about. Just sit down,' because I'm standing in the middle of the dining room. So everybody sat down. And a busboy ... ran around the corner of the fireplace and said to me, 'It's not all right. The place is on fire.' I said, 'No, it's not, everybody be seated.'*

*He said, 'No, ... said the place is on fire—get everybody out.' And I still thought he didn't know what he was talking about because I've worked there too long ...*

*So people jumped up again and said, 'The place is on fire.' So I said, 'Okay then, walk out, don't run.'*

*... They started toward the back. They said, 'There's an exit.' I said, 'No, you can't go that way because it's a kitchen and it's dangerous to go through the kitchen because of the floor ... because the floor is slippery, if you're not careful, you're gonna fall, it's gonna be our fault.' So when they said, 'We can't go through the front,' I said, 'Okay, everybody go through that door, follow that door, go straight on through, back through the kitchen, there's an exit at the end of the kitchen where you can get out. Don't panic, just walk.' Because I still thought it was just smoke. So I started through and all the customers were in front of me.*

Even as word of the fire was spreading, additional customers were arriving and entering the building to have dinner and then enjoy John Davidson's 10:30 show in the Cabaret Room.

Heavy smoke in some areas forced people away from the nearest exits. They had to find their way out through unfamiliar sections of the building. One patron in a group of 22 stated:

*... We had just sat down. This was approximately 8:45. We had just sat down and the waitress had come around and asked us if*

# 1. THE BEVERLY HILLS SUPPER CLUB FIRE

we wanted another before-dinner drink. *The people were ordering their drink, and the waitress had left to go get our drinks, and I would imagine in a matter of minutes, we hadn't even gotten seated properly where we were gonna sit, and a young fellow came in and said, 'Ladies and gentlemen, would you please leave the building, we have a slight fire.' We all proceeded to get up. I was sitting at the end of the table closest to the lounge. We got up and a waitress ... said, 'No, it's not that much, please sit down.' So we all went and sat down again. In a matter of minutes later again, the same young fellow came back and said, 'No, you're going to have to leave. We're going to have to evacuate the building.' So I grabbed my wife by the hand and we proceeded to leave through the lounge area, the way we came in, to go through the main entrance. We went through the dining room and got to the lounge area, and the smoke was so intense my wife started coughing and saying, 'I can't go any further.' Well, that was the only way that I knew out. I had been to Beverly Hills several times, and I said, 'We got to go this way.' So we went through the smoke up to where the lady stands and checks you in for your reservations, and we could go no further cause there was too many people. They were just crowding out that door. My wife and I couldn't take the smoke so we turned around, and my party was following me, and I said, 'We can't get out this way, just turn around and go back.' So we went back in the main dining room.*

*... The main dining room had no smoke at all, maybe a mist, but no heavy stuff.*

*... one of the head girls ... was standing in the back saying 'come on, go out this way.' There was an exit through the kitchen. So by that time, my whole party had gone and my wife and I were the last ones. In fact, before I went out the door, I said to the (the hostess), 'Come on, get the hell out of here.' She said, 'No, I am not leaving until I make sure everybody is out of here.' I said, '... everybody is*

out of here, we're the last ones.' She says, 'No, there might be some more people coming through here.'

Another patron related:

> ... Someone said something about a fire. We heard people say maybe it was just a joke. They asked if it was a fire drill. The next thing we knew all these people were running down the hallway to the exit beside the Cabaret Room. We started walking out slowly because we thought this was a fire practice. We heard a maître d' say, 'Yes, it is a real fire get out.' Then he started to tell everybody to get out. We walked past the Cabaret Room and they were all still sitting in there and the comedians were still going with their show. They hadn't even bothered to get out.

The people in the Cabaret Room were first alerted at about 9:06. A busboy came in and made his way to the stage. One of the comedians handed him a microphone. He showed the patrons where the exits were, told them there was a small fire elsewhere in the building, and instructed them to leave. Some patrons thought it was a joke, and many remained in their seats. Some started to leave. One patron recalled:

> ... And these comedians were performing. They were just hilarious, and we were relaxed and enjoying it, when this busboy just came on the stage, it just didn't register with people. That's all I can say. I think the comedians said something to the effect, but people just weren't moving, and they said something to the effect that this was serious, that we'll come back in a few minutes and we'll continue where we left off.

As Cabaret Room patrons started to exit, smoke filled the passageways,

## 1. THE BEVERLY HILLS SUPPER CLUB FIRE

advancing in some places more rapidly than anyone could have imagined. Panic set in at some locations, particularly where heavy smoke and flame suddenly appeared. One of the surviving musicians recalled:

> *... And we quickly gathered our belongings went out that door and we went down the sidewalk along the side of the Viennese Room and we walked a few steps to our left down on the grassy hill there and turned around to see who else would come out. We weren't there very long ... I would say not more than a minute or two, probably not that long. A huge ball of flame came out the door, and then smoke just before that, black smoke that everybody talks about. And after that I would say only a few people came through that door and then there were no more people who came out under their own power.*

The margin of safety for those who escaped was often very thin. One waitress recalled:

> *... I wasn't thinking about anything but getting out of there, because the smoke was right on us. I got up to the double door, which did not lead outside, it led to another bar ... and we surged backwards and we surged forwards and when we did, I grabbed a hold (of) a man's collar. He pulled me through and I turned around to look because the smoke and the flames were coming, the smoke was coming out of the double doors out of the Cabaret with me, right at the back of my head. There must have been flame, because my blouse was burnt. As I came out, I turned to look and the people weren't screaming any more. The smoke had covered them all up.*

Firemen approaching from the highway saw smoke coming out of the eaves of the building along its entire length. Arriving at the main entrance, they encountered people streaming out. The firemen had to push and shove their way in to assess the situation. Once inside, a fast-growing fog of thick smoke prevented them from getting very far.

More emergency equipment arrived. Hoses were set up and fire fighters reacted to information coming from occupants who had been in different parts of the building. Early on, the firemen could not determine where the fire was, and fire suppression initiatives were futile. Hose lines at the main entrance were quickly abandoned when the focus shifted to rescuing occupants from the Cabaret Room at the opposite end of the building. The National Fire Protection Association (NFPA) report describes the activity at that time as follows:

> *At the front of the club (south side), one of the owners recognized a Southgate fire lieutenant, grabbed him, and said 'My God, we've got a situation around the side.' The lieutenant led a crew of fire fighters to the east side. Other reports were received at the same time that people were trapped in the Viennese Rooms and that the fire was into the bar and off to the right. Fire fighters took 1-1/2 inch and 2-1/2 inch hose lines in the front door, and almost immediately a hysterical report was received from an employee in a tuxedo that people were trapped in the Cabaret Room.*
>
> *The chief ordered that the lines in front be abandoned, and all available fire fighters then went around to the Cabaret Room to assist in rescue operations. The lieutenant who had taken a crew around the east side was at the chapel exit ... and was coordinating efforts to drag victims from the building. Employees had already started pulling people out before the fire fighters arrived to help.*
>
> *Victims were being dragged out bodily. One fire fighter reported that he put breathing apparatus on and went in as far as he could get. However, he couldn't enter through the service doors because people were stacked in the doorway. Other victims were at the service bar, and others were lying on the floor in the hallway. Rescue*

# 1. THE BEVERLY HILLS SUPPER CLUB FIRE

*operations became confused as a result of too many people getting in the way; finally the rescue was organized, with seven or eight fire fighters pulling victims to the exit door, and employees taking these victims from the exit door away from the building.*

The fire fighters had come around to that part of the building at about 9:15, less than ten minutes after the Cabaret Room patrons had first been alerted, and less than 15 minutes after the first call to the Southgate fire department. Doctors and nurses among the patrons were already doing their best to help those victims who had been pulled out. They helped those who were still alive and they succeeded in resuscitating some who appeared to be dead. The grounds around the club became a grim collection of the dead, the injured, fire and rescue personnel, and patrons wandering around trying to find their relatives and friends.

Meanwhile, the fire continued to spread. There was little chance of survival for people still trapped inside. Rescuers continued to enter the building, attempting to locate and remove victims. They were often unable to pull living people out of the pileups at some exits. An employee recalled:

*... there was no question in my mind about how serious this thing was at that point because people were just stacked up and they were alive you know. Maybe they still are I don't know but there were maybe five or six bartenders that I worked with at one time or another and they were getting as many as they could get out of there.*

*... there was a woman ... and she was out of this thing except there was about six people, there was enough people where we couldn't get her out at this time because they were on her legs, the back of her legs. She was clear of this thing but she had all this weight on her and we just about pulled her eyes (sic) out of their sockets and we couldn't move her. And then there was a man that was (on) top, and he was reaching his arms up and so I thought he was all right and he looked*

*like he was on top and that was the first thought, get him off top so you can do something with the bottom ones and anyway I had him wrap his arms around my neck and I pushed up against this door as hard as I could and I moved the guy about this far, about two feet and about this time he was out (of) it, he didn't have much strength left to help me and I didn't have enough strength to lift him and he just looked at me and shook his head you know, there was nothing I could do and then there was a young girl, she was (on) top and she was alive and well and everything else, she wasn't screaming or anything but she was in fine shape, and I started to walk out with her and her leg was wrapped around a table, I don't know how that table got there but there was a table leg and it was just entangled in there and she couldn't pull it loose. So that's my recollection, I can't tell you the one (sic) I left there, I couldn't take any more, That's I think when I left there, it was terrible, just terrible.*

Eventually, despite the best efforts of hundreds of fire fighters from Campbell and surrounding counties, the fire was clearly out of control and the interior of the Club became too dangerous for further rescue attempts. According to the NFPA report:

*At about 11:30 p.m., it had become apparent that the fire departments could do nothing inside, and fire fighters might be injured or killed. The decision to evacuate personnel from the building was made when the fire was completely out of control; the fire was not under control until approximately 2 a.m. The search for the remaining victims was begun at daylight the following morning, although the fire was not completely extinguished until Monday Morning, May 30, 1977.*

There were 161 fatalities at the site and many injured, some of whom died afterwards in hospitals. At the time the NFPA report was written, the death toll from this fire was 164 people. The final total was slightly higher. Thirty-four of the victims were severely burned, most likely after they had expired from smoke inhalation.

Before the shock, grief and mourning had subsided, a sense of outrage started to build as investigations revealed serious safety deficiencies in the construction of the club, its operation, and in the building permit and inspection process. There were investigations, inquiries and hearings delving into the cause of the fire, its rapid spread, fire safety deficiencies in the construction of the building, the operation of the Club, the overcrowding of the Cabaret Room, and the regulatory inspectors who seemed to have sidestepped their responsibility for public safety.

The official investigations identified the wall or ceiling space of the Zebra Room as the area of origin. Investigators concluded that the most likely source of ignition was an electrical malfunction, the exact nature of which they could not determine. A grand jury determined that acts of negligence had contributed to the tragedy, but no criminal indictments resulted.

Beverly Hills Supper Club—The elegant spiral staircase after the fire. (photo: J. Aronstein) (Visit *aluminumwire.info* for links to Supper Club photos before, during, and after the fire.)

Civil lawsuits were initiated by victims' families. Most were consolidated to proceed as class action lawsuits in various categories. The Club's owners, the power utility, and governmental agencies were sued on the basis of negligence. Overcrowding of the Cabaret Room, inadequate exits, failure to perform required inspections, and a multitude of other violations of standard and required fire safety provisions formed the basis of these negligence claims. By about 1980, almost all these lawsuits had been resolved. Governmental defendants escaped liability for failure to enforce existing standards by successfully claiming *sovereign immunity*.

Product liability lawsuits were filed against manufacturers of the furnishings and decorating materials in the club that were the source of the thick black toxic smoke. Plaintiffs charged that these highly combustible plastic products were intrinsically hazardous in a fire situation and directly

# 1. THE BEVERLY HILLS SUPPER CLUB FIRE

contributed to the fast spread of the fire and loss of life. This group of product manufacturers eventually settled, some right after a jury found for the plaintiffs against the producers of PVC plastic materials.

Plaintiffs' attorneys focused on a possible cause of the fire—an aluminum wire connection failure. During a renovation in the early 1970's, some aluminum wiring had been installed in the vicinity of the club's Zebra Room to feed receptacles and lighting.

By 1977, when the fire occurred, the safety issues associated with aluminum wire were well known. The U.S. Consumer Product Safety Commission (CPSC) had declared it to be a hazardous product. The agency initiated legal action against the aluminum wire and wiring device manufacturers, seeking to have them alert the public to the hazards and pay for corrective measures. The CPSC's action was based on an unusually high incidence of aluminum-wired connection burnouts in homes across the country, some of which resulted in fires, injuries, and deaths.

Arguing that an overheating aluminum wired connection was the most probable cause of the Beverly Hills Supper Club fire, plaintiffs' attorneys took legal action against the producers of the aluminum wire, the manufacturers of the wiring devices that were sold to be connected to it, and Underwriters Laboratories Inc. (UL), the respected testing lab that had vouched for its safety.

# 2.
# FATAL FIRE—HAMPTON BAYS, NEW YORK

*On April 28, 1974, two people died in a house fire in Hampton Bays, N.Y. Fire officials determined that the cause was an overheating aluminum wire connection at a wall receptacle.*
(CPSC Publication #516)

THE *CAUSE AND ORIGIN* INVESTIGATION OF A FIRE GENERALLY STARTS BY determining where a fire started, which is the "area of origin" or "point of origin", followed by an analysis of how it started, which is the "cause". The process is often thwarted by the amount of damage inflicted by the fire itself and by rescue and fire suppression operations disturbing the physical evidence that remains. From the eyewitness accounts and the site photo previously shown, it is easy to understand why the exact point of origin and cause of the Beverly Hills Supper Club fire could never be determined with anything near 100% certainty.

In contrast, there was little structural damage from an earlier fire at the Hersh residence in Hampton Bays, New York. The point of origin and the cause were precisely determined. Investigators agreed that ignition occurred as a result of an overheating aluminum wire connection on a receptacle. There was no challenge to that conclusion. For that reason, the Hersh fire still serves today as the CPSC's prime example of the aluminum wire hazard.

Hampton Bays is on Long Island, NY, about 50 miles east of Melville, where UL has a major facility. UL is viewed by the public as the gatekeeper of electrical safety. Its logo on a product signals to everyone involved—distributors, electricians, inspectors, and the general public—that UL has examined and tested the product and considers it to be suitable for its intended purpose. The UL label implies that the product will perform safely. When the Hersh home was built, the aluminum wire and other electrical system components could not have been marketed, installed, and then approved by inspectors without having the UL label.

The Hersh home had already been occupied for two years before UL started substantive testing of aluminum-wired receptacles. Previously, for about a decade, UL had repeatedly advised the trade and the public that the terminals on all UL listed receptacles were suitable for use with aluminum wire. However, UL had not done any testing to determine if that was correct.

Copper wire had long been the standard for residential electrical systems. Aluminum branch circuit wire was introduced into the US housing construction marketplace in 1965. Connection problem reports started to surface soon afterwards. UL eventually initiated a test program in 1970, in response to a rising tide of reported connection burnouts and fires in aluminum wired homes.

The Hersh family's home was a 3-bedroom raised ranch, completed in April 1968. All of the original wiring was aluminum. The fire started inside the wall at a receptacle in an unoccupied bedroom. It most likely had smoldered and burned for some time before the smell of smoke woke Mrs. Hersh. As she recalled later in testimony at a Congressional Committee hearing:

> *On April 28, 1974, in the early morning hours, my husband, daughter, and I were asleep. My sons were not at home. I was awakened by the smell of smoke. I got out of bed and saw the smoke coming from the small bedroom across the hall, which was in flames. I awakened my husband and I ran to my daughter's bedroom to awaken her. The blaze was too intense and I was forced back. I then ran into my*

## 2. FATAL FIRE—HAMPTON BAYS, NEW YORK

*bedroom and tried to call the fire department on the telephone but couldn't get through ... and collapsed on the bed.*

Investigators surmised that Mr. Hersh went into the hallway, was quickly overcome by the smoke and collapsed. Their daughter Sheri awoke, opened her bedroom door and encountered smoke and flame in the hallway. She then retreated and collapsed on the floor near the window in her room.

A fireman living nearby was first on the scene. He entered the home intent on rescuing occupants, but was forced out by heavy smoke and heat, and later collapsed. The three people in the house were finally brought out, and resuscitation efforts began. Only Rosalind Hersh responded. Her husband Jack and their daughter Sheri died.

The fire hazard posed by overheating aluminum wired receptacle terminals is readily appreciated by inspecting field failure samples, such as the one shown in the photo at right. During normal use in the home, this receptacle's plastic body and wire insulation disintegrated from heat generated at its aluminum wire connections. With the receptacle's cover plate in place, there is nothing visible from the front that would alert occupants of the home to the developing hazard. Occupants occasionally detected overheating receptacles by a burning smell or an abnormally hot feel. An unusually high number of incidents of overheating aluminum wired receptacles, in various stages of deterioration and hazard level, were being noticed and reported across the country. It was unlike anything that had ever been experienced with copper wired receptacles.

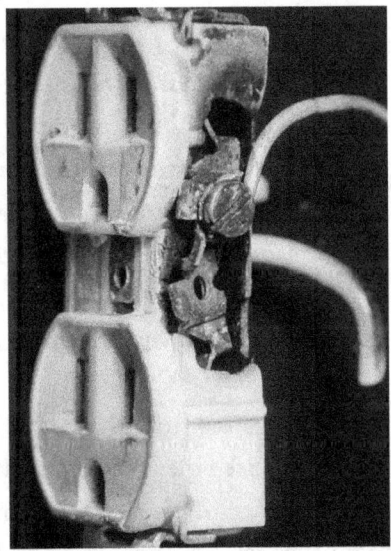

*Photo: J. Aronstein*

Receptacle circuits are typically wired in "daisy chain" fashion. Current flowing from the service panel to an item plugged into to a receptacle far down in the chain may pass through two dozen or so wire terminations on the intervening receptacles. Any deteriorated wire connection in the current-carrying part of the chain may overheat when current is flowing, even if nothing is plugged in to its particular receptacle.

At the Hersh home, nothing was plugged into the guest room receptacle at the point of fire origin. The current that caused one or more of its aluminum wire terminations to overheat passed through the wiring in the walls to a combination humidifier/dehumidifier plugged into a receptacle in the hallway. It drew 4.8 amps current when it was on, only one-third of the circuit rating. Subsequent to the fire, a laboratory test using an identical unit demonstrated that the heat generated at a failing aluminum wired receptacle terminal at that seemingly low current was sufficient to produce fire ignition conditions.

Mrs. Hersh testified that:

> *(The house) ... was the most important and costly purchase of our lives and I guess we had a right to assume that something that had been around as long as electricity would be supplied to us safely. But a time bomb had been built into our house and we didn't know it. The electrical system had a weak link—aluminum wire. Nothing happened until sometime in 1972 when our dog began sniffing around a wall receptacle in the dining room. My husband discovered that it was warm to the touch and replaced it.*

Investigators from the CPSC and the National Bureau of Standards (NBS) inspected the Hersh home after the fire and found several receptacles with evidence of overheating. Among them was the one in the dining room that replaced the warm receptacle that the dog had been sniffing. Connections on the replacement receptacle were overheating only two years after it was installed. The Hersh family did not know that the overheating receptacle which they found and replaced indicated a systemic problem. Mrs. Hersh testified that:

> *Nobody ever told us that aluminum wire was dangerous. I know now that the aluminum industry and Underwriters Laboratories were aware long before this tragedy that there was something very wrong with aluminum wire; that there were fires all over the country in homes wired with aluminum and yet they did nothing about it.*

Local awareness of aluminum wiring and its potential hazards quickly spread as a result of the Hersh fire and related publicity. At a later time, the NY State Director of Product Safety noted that, after the Hersh fire, "… following local press release, an estimated 400 calls to the CPSC from neighboring areas allegedly reported similar episodes." Before they entered the Hersh home on their first day there, the CPSC and NBS investigators were shown a set of burned splicing connectors from a neighbor's aluminum-wired range circuit.

The Hersh family sued Kaiser Aluminum and Chemical Corporation, the manufacturer of the KA-FLEX brand of aluminum wire non-metallic (NM) cable that had been installed in their home. Maury J. Rubin, an attorney with an office in NY City, represented the Hersh family. The prestigious firm of Sullivan and Cromwell represented Kaiser. This was a David and Goliath contrast. Sullivan and Cromwell occupied two or more richly appointed floors of a downtown NY City skyscraper with a commanding view of the harbor. One could not help but be awed by the offices of this century-old firm. They were furnished with massive solid wood conference tables and desks, and a century's worth of founders and partners looked down from large oil painted portraits hung on the walls. Early in the 20th century, one of the firm's founders, William N. Cromwell, had been heavily involved in deflecting congressional action away from construction of a canal through Nicaragua, so as to salvage—for his own profit and for the enrichment of J.P. Morgan and others—the failed French company's rights to a canal through the part of Columbia that is now Panama. He was then involved in the somewhat shady history of the birth of the Republic of Panama.

At the Sullivan and Cromwell office, Rubin took the deposition testimony of T. R. Pritchett, manager of Kaiser's research and development laboratory. Pritchett testified that the product in question, branch circuit aluminum wire, had been thoroughly tested. This would generally be taken to mean not only that it was tested, but that it also performed well in the tests. Information developed after Pritchett's deposition showed that Kaiser did indeed do extensive testing, most of it after the Hersh fire, but the results were generally poor. The Kaiser testing consistently showed that the product that they had been marketing was not compatible with many types of common wire terminals and splicing connectors that had to be used with it to wire a house.

In 1965, Kaiser started marketing KA-FLEX aluminum NM cable for branch circuits in homes. Contrary to all previous experience, test results, and trade practice, the company promoted it to be installed just as copper wire would be, without the need for special installation procedures or special connectors. Within Kaiser, engineering knowledge and concerns had yielded to marketing imperatives, setting the stage for tragic incidents such as the Hersh fire. Kaiser's marketing rationale was reduced to this: if the product was allowed by the National Electrical Code (NEC), was UL listed, and the electrical inspectors approved the installations, then nothing more was required. The company's objective was to sell aluminum.

Kaiser's strategy could not be well defended in front of a jury. The Hersh v. Kaiser litigation settled for an undisclosed amount without going to trial.

# 3.
# KAISER ALUMINUM AND CHEMICAL CORP.

Henry J. Kaiser was a well-regarded industrialist, known for construction, shipbuilding and steel. Before the Second World War he was involved in building some of the largest dams in the country. During the war, using innovative techniques, Kaiser transformed shipbuilding the way Henry Ford had transformed automobile production a half century earlier. His shipyards produced more than 1,400 cargo ships and 50 small aircraft carriers. After the war, Kaiser's companies redirected their production capacity to satisfy peacetime markets, producing products ranging from dishwashers to automobiles. Today, his best-known legacy may be the Kaiser Permanente health care network, which grew out of the health care systems that he developed for his employees.

Two government-owned aluminum plants were built in 1943 near Spokane, Washington to help meet aircraft production needs. The Mead plant was a smelter, which produced the aluminum metal. The Trentwood plant, about ten miles away, was a rolling mill that created aluminum sheets from metal produced at the Mead plant. The two factories were located inland for protection against Japanese attack, and were separated to provide an increased level of protection.

They went up for auction as surplus in 1946, and were purchased by Kaiser Permanente Metals Corporation. In 1949 the company name changed to Kaiser Aluminum and Chemical Corp. That same year, it acquired an aluminum bar, wire and cable mill in Newark, Ohio, which put it in the electrical wire and cable business. Aluminum was increasingly being used as an economical choice for outdoor electrical transmission and distribution lines, the ones that stretch from tower to tower and pole to pole.

By then it was well known that connectors successfully used for decades with copper wire and cable could not be applied to aluminum with any assurance of long-term failure-free service. The underlying causes of aluminum connection failure were not well understood at the time. Consequently, the industry's testing methods could not reliably sort aluminum connection types that would fail in long-term service from those that would not. Efforts toward solving the problem by trying new tests and developing connectors specifically designed for aluminum were very slow in yielding meaningful results.

In promoting the use of aluminum wire and cable, the manufacturers commonly point to the many miles of aluminum outdoor transmission lines that have been in service for years. The implication is that if aluminum has been selected and used by the power utilities, and it carries current at thousands of amps and hundreds of thousands of volts through the harsh outdoors from the hydroelectric generators at Niagara Falls to New York City, then surely you can depend on it to be safe in your home or commercial building at far lower current and voltage.

The use of aluminum for outdoor transmission and distribution lines has been an economic success, but not a complete technical success. The economic balance sheet for the power companies includes the cost of dealing with a persistent problem of connection failures where the aluminum cable is terminated, spliced, or tapped onto. The ongoing aluminum connection problem in transmission and distribution lines was reflected early on in the title of a 1952 article by an engineer with Pacific Gas & Electric (PG&E), "Better Connector Life Vital to the Use of Aluminum in Distribution".

Transmission line operators found it necessary to perform periodic inspections using aircraft equipped with infra-red cameras to identify hot connections. Today, unmanned drones are also used for that purpose. Failing

# 3. KAISER ALUMINUM AND CHEMICAL CORP.

connections were identified and serviced or replaced. The cost of continuous inspection and maintenance of connections is more than offset by the savings gained by the lower price of aluminum cable relative to copper. For the electric power utility companies, hot connections on the outdoor lines are a manageable nuisance.

The situation for wiring in buildings is quite different. There are no tiny helicopters or drones that can flit around inside the concealed and remote spaces in buildings to find overheating connections and send alerts to the occupants. The consequence of an overheating connection failure inside a building's walls is, in the worst case, a fire. That was well known in 1945, when WWII ended.

Except for war emergency situations and some experiments, copper was universally used for the electrical wiring inside buildings. However, copper was in short supply in the US marketplace during World War II and for a time afterwards. It was classified as a strategic material. In contrast, at the end of the war, aluminum was plentiful. An excess of aluminum production capacity developed as wartime aircraft production wound down.

Until then, the UL standard for insulated building wire recognized the use of copper only. At the end of the war, with copper scarce and aluminum available, there was an exploding demand for new housing. UL was pressured by industry and government to accept the use of insulated aluminum wire and cable for use inside buildings. The aluminum manufacturers wanted to keep their production facilities operating, and the government was trying to ward off a shortage of housing for returning war veterans who would soon be starting their own families. The government anticipated what later became known as the postwar "baby boom".

When UL bowed to the pressure in August, 1946, it accepted insulated aluminum building wire on a temporary emergency basis. Alvah Small, UL President at the time, wrote a press release plainly stating that the temporary acceptance was necessary to accommodate an emergency shortage of copper, and that it most likely would "add to the hazards of life." Within UL at that time, it was known that aluminum wire was not compatible with connection types that were used for wiring in buildings—specifically including the screw terminals on receptacles.

Almost immediately, U.S. Rubber, a wire and cable manufacturer, requested permanent recognition of aluminum building wire. But UL did not budge from its position of allowing aluminum only on a temporary emergency basis for several years, citing the need for thorough testing. Connector tests by UL and others at that time showed that aluminum conductors performed poorly relative to copper. Overheating of connections inside buildings could not be tolerated. It was a known fire hazard.

As a manufacturer of insulated wire and cable, US Rubber bought the metal that it used, and shipped out wire with the rubber or plastic insulation and cable jackets added. The company had no stake in the selection of aluminum or copper conductor; its customers determined what they preferred. The 1950 US Rubber catalog of their complete insulated wire and cable offerings contains only a small selection of aluminum wire products relative to the copper wire products. UL's temporary approval for aluminum was a damper on aluminum sales.

In 1951, UL accepted aluminum building wire on a permanent basis. Connection test results were still poor, but the requirement that the wire should perform well in connections was cast aside. UL's previous position reflected a concern for safety. It is fundamental that electrical fire safety in buildings hinges on the wire and connectors working properly together as a system. UL's new position was that wire and connectors were separate products and could be evaluated independently. UL subjected the aluminum wire and its insulation to various tests, but no longer tested its performance at connections. The aluminum wire did not have to perform safely at splices and terminals to gain UL's approval.

Aluminum building wire sales still did not take off. Even with permanent recognition by UL, the reputation that aluminum conductor had for poor performance in connections continued to depress sales. Kaiser

# 3. KAISER ALUMINUM AND CHEMICAL CORP.

personnel recognized that there was a connection problem, but laid the blame on the connectors. From their point of view, the aluminum conductor worked just fine to carry the electricity from one point to another, but the connectors were inadequate. Most others in the industry cited the intrinsic properties of the aluminum metal as the cause of the connection failures.

At that time, several properties of aluminum were known to reduce its connectability relative to copper. Most important is that a transparent hard film of aluminum oxide immediately forms on any aluminum surface exposed to air. Aluminum oxide is a very effective electrical insulator. In contrast, the surface films that form on copper are slow growing and electrically conductive.

In an electrical connection, where bare aluminum is pressed tightly against another metal surface, virtually all of the physical contact area is electrically insulated by the aluminum oxide unless it has somehow been mechanically displaced. When Kaiser entered the electrical wire and cable business in 1949, the accepted installation practice for aluminum was to abrade the wire surface with emery cloth or a wire brush. This removed the oxide layer, exposing a fresh metal surface. This was commonly done under a coating of some sort of grease to keep the surrounding air from forming fresh oxide. Inhibitor grease was also applied to the connector surfaces that would contact the aluminum conductor.

Compared with installing copper wire and cable, installation of aluminum with abrasion and inhibitor is time-consuming and somewhat messy. The added cost was factored into the economics of using aluminum conductors. Until 1965, when Kaiser introduced its KA-FLEX NM cable for residential wiring, the company's own installation instructions and advice to customers included—and forcefully emphasized—the need for abrasion and inhibitor.

Even with the use of abrasion and inhibitor, the failure rate for aluminum conductor connections far exceeded that for copper conductor connections. The connection/connector problems weighed heavily on Kaiser's aluminum wire and cable sales. In 1952, Kaiser embarked on a program of applying its own test procedures and qualification criteria to develop lists of terminals and connectors that could be recommended to their customers with confidence. UL listing alone was not enough to assure satisfactory and safe long-term connection performance. The UL tests for connectors and terminals suitable for use with aluminum conductor were not sufficiently comprehensive and stringent. Kaiser also established a "Connector Propaganda Committee" to develop ways to counter the adverse publicity that aluminum wire and cable was receiving and to shift the blame for burnouts to the connector manufacturers.

By 1957, Kaiser's connection testing program was well under way. It revealed many problems. For example, one of the first applications for which aluminum conductor had been commonly used in or on buildings was service entrance cable. This is the cable that brings the power to the home from the utility company's overhead line or transformer and brings it to terminals in the meter base, where the electric power meter is mounted on the building.

In 1949, when US Rubber first sold aluminum service entrance cable, there were many burnouts of meter base terminations. Eight years later, in 1957, Kaiser's C.G. Sorflatten warned that if aluminum cable was used in the meter base connectors available at the time, the burnout problem would emerge again.

His warning was firmly supported by his test results. In one test, Sorflatten applied rated current, 200 amps, to an aluminum-wired Walker meter terminal. It failed on the first cycle at its rated current, heating to more than 500 F. On its second current cycle, again at its rated current, the insulation on the aluminum cable started to burn. In contrast, a Duncan meter terminal performed well in the same test and could be recommended by Kaiser.

In 1957, Kaiser acquired US Rubber's wire and cable plant in Bristol, Rhode Island. Under the new ownership, the plant's previous metal-neutral

## 3. KAISER ALUMINUM AND CHEMICAL CORP.

sales approach changed. Instead of providing wire and cable with either copper or aluminum conductors, whichever the customers wanted, the sales force was directed to push sales of aluminum. The purpose was to consume and profit from the metal that Kaiser's own smelters and mills could produce. The Bristol plant and its associated sales organization then actively promoted aluminum building wire products, even if the profit margin was lower than for copper wire, and even though it was known that the expected reliability in connections—and therefore its safety—was inferior to copper.

Kaiser's marketing objective was to displace copper in all electrical wiring and equipment applications. The company envisioned a diminishing supply of copper that would make the transition inevitable. In 1956, Kaiser started to focus on indoor residential wiring applications. As a first step, some aluminum wiring was installed in an employee's house. Kaiser's G.N. Houck wrote:

*We are going to use Howard as a guinea pig to try out aluminum building wire in several applications in his house.*

At about the same time, Kaiser's C.G. Sorflatten had been reaching out to the connector and wiring device manufacturers. He offered assistance toward making their products compatible with aluminum wire and cable. Very few of the manufacturers that he approached shared the Kaiser vision of aluminum taking over the wiring world. They were reluctant to make changes, at their expense, to products that were already on the market and working just fine with copper wire.

For Kaiser, the connection problem remained a major obstacle to acceptance of aluminum wiring of all types. There was confusion in the electrical equipment marketplace as to which connectors or equipment could be used with aluminum, which would be accepted by the local inspectors, and most important, which would actually be safe over the long term? Aluminum wire sales did not meet expectations.

Within Kaiser, Sorflatten's rigorous connector test and certification program was under attack. It was holding back sales. For example, Sorflatten tested range and dryer receptacles. None met Kaiser's in-house performance requirements. The field salesmen could not point to a Kaiser-recommended range receptacle for use with aluminum cable. They had to discourage the sale of aluminum wire for that application.

Kaiser's J.P. Moran, summarizing the company's aluminum connection problems and programs in September 1957, noted that the company's position was to sell aluminum for use with existing terminations, "taking a calculated risk", while simultaneously engineering to lower the risk.

At two other Kaiser labs (not Sorflaten's), tests were being conducted to determine if various connectors on the market, either as-is or with minor modifications, met UL requirements for use with aluminum. The need for Kaiser's more stringent performance standards was being challenged by those who thought that the tests performed by UL were sufficient. Sorflatten defended his rigorous approach. He argued that the UL test requirements were known to be inadequate and serious fires could result from connection failures in buildings. Sorflaten lost the debate. Connector testing and research at the various Kaiser labs, including Sorflatten's, was eventually put under the control of W.K. Priestly, who did not support Sorflatten's conservative safety-first approach. About a year later, Sorflatten left Kaiser for a position at ILSCO, a connector manufacturer.

Earlier, Sorflatten had been involved in evaluating and selecting wiring devices for use at the Colonial Village housing project in Ravenswood, West Virginia, where a Kaiser plant was a major factor in the local economy. The project was to be used as a trial installation of aluminum branch circuit wiring. It would later become the centerpiece of Kaiser's promotional claims that their residential aluminum wiring product, "KA-FLEX", had been thoroughly tested.

The Ravenswood project consisted of 2-story buildings with a total of 94 apartments. It is a type of development sometimes called "garden

# 3. KAISER ALUMINUM AND CHEMICAL CORP.

apartments". The first units constructed were scheduled to be wired in mid-April, 1958. Sorflatten tested wall outlets and range receptacles from various manufacturers to select those that performed well with aluminum wire for use at Ravenswood. Initially, none that he tested met the Kaiser requirements. Eventually, wiring devices that performed well with aluminum wire were selected and installed. The installation was done according to Kaiser's recommended practice of the time, with abrasion of the wire and application of corrosion inhibitor compound.

UL had not done any testing to qualify the available receptacles and switches for use with aluminum wire. In early 1958, when construction started on the first Ravenswood units, UL clarified its practice as to markings on equipment and fittings for use with aluminum conductor. Noting that there was much misunderstanding, UL stated that equipment and connectors could be considered to be suitable for use with aluminum only if marked "AL" and "CU". Otherwise, the equipment could only be used with copper conductor. None of the receptacles and switches on the market had the AL-CU marking that inspectors were told to look for. This was a roadblock to local electrical inspectors' approval of aluminum wire installations.

UL provided a letter to cover this situation. Without having done any connection performance testing to back it up, the letter stated that all UL listed wiring devices were considered to be suitable for use with either copper or aluminum wire. This served Kaiser's needs at the time, and the Ravenswood installation passed inspection. For the next 14 years or so, still without having done any supporting tests, UL continued to send out versions of the same letter in response to inquiries as to the use of wiring devices with aluminum wire.

Kaiser's R.P. Brown visited the Ravenswood complex five years after the first apartment was wired. His report indicates that he was told about burnouts of aluminum wire splices in ceiling fixtures. Additionally, based on his own inspection, Brown predicted that there would be problems with the aluminum service entrance cable connections sometime in the future, but he suggested that the power company would most likely be blamed, not the aluminum cable.

A few weeks later, Kaiser's H.H. Borup visited Ravenswood and inspected the wiring in five apartments. He found an overheating aluminum wire termination in one of the circuit breaker panels. He took the position that it had been improperly made and therefore was not "chargeable" to the aluminum conductor. Borup wrote that he thought the aluminum wiring had performed as well as copper would have.

Kaiser cited the Ravenswood project and their one employee's house as being their successful tests of aluminum house wiring. In July, 1965, just after the company introduced KA-FLEX for residential construction, the company started a series of ads in IAEI News, the monthly magazine of the International Association of Electrical Inspectors. In its ads, Kaiser claimed that these wiring projects had given it 10 years of experience with aluminum house wiring, that the connections were all as good as new when inspected, and that UL had done extensive research on aluminum wire connected to receptacles and switches in 1952. These claims were not true.

The objective of the 1965 Kaiser advertising campaign in IAEI News was to overcome negative attitudes about aluminum wiring. The inspectors had to be convinced. Kaiser's ads and promotional material did not disclose that in fact there had been aluminum wire connection failures at Ravenswood, that the wiring devices used there had been specially selected, and that the connections had been made using wire abrasion and inhibitor compound.

In 1979, R.T. Noonan, an attorney with the U.S. Consumer Product Safety Commission (CPSC), visited the project and reviewed the local fire department's records. He found that in the relatively short time (20 years) since the apartment complex had been completed, there had been four fires in the complex reported to be caused by aluminum wire connection failures. CPSC personnel also documented additional serious incidents of aluminum wire connection failures at Ravenswood that did not result in fire. This was

## 3. KAISER ALUMINUM AND CHEMICAL CORP.

a very poor safety record for residential wiring, especially considering the special care that went into the installation.

The market potential of aluminum building wire varies with the relative prices of copper and aluminum metal. If aluminum wire is the same price as copper, contractors will favor copper wire. If aluminum is priced lower than copper, the aluminum wire becomes attractive, all the more so when a few dollars difference can win a bid. Kaiser had been poised to enter the residential NM cable market since about 1958, but the aluminum-copper price see-saw did not tip sufficiently to make it worthwhile. That changed in 1964, when the presidential election in Chile impacted copper prices in the USA.

That election pitted the Socialist party's Salvatore Allende against the Christian Democratic Party's Eduardo Frei Montalva. Allende's platform included complete nationalization of Chile's North-American owned copper mines, which produced a significant portion of the world's supply. It was a concept that appealed to the voters. Frei followed suit with a somewhat milder proposal, calling instead for Chile to take only a controlling majority interest of the North American mine ownership. Either way, the US copper companies would likely be hurt financially.

With the help of covert funding from the US CIA, Frei, the more appealing candidate to the North American interests, won the election. The events in Chile, along with increasing demand for copper in the U.S. due to the healthy economy, the "space race", and the start of the Vietnam War, caused a jump in the price of copper. The U.S. government released copper from its reserves in order to satisfy the demand and stabilize the price. But the price was high enough to make aluminum NM cable an economically viable product for the residential housing market, especially where lowest bids won the contracts for large scale projects.

Lower cost alone could not make aluminum NM cable a clear winner if special installation procedures and special wiring devices were required to install it. Kaiser re-wrote its installation instructions and promoted the concept that it could be installed just the same as copper wire.

At the beginning of 1965, the company started producing and marketing its KA-FLEX aluminum NM cable. Kaiser was the first company to market this product for residential use. The company took the industry by surprise, and opened a Pandora's box of electrical fire safety problems.

# 4.
# FATAL FIRE—COLUMBUS, GEORGIA

Peggy Johansen put 15-month old Janet to bed in her crib at about 10:00 p.m. on the day after Christmas, 1978. A little later, she put Janet's 4-year old brother to bed in his room, turning on a vaporizer and a night light just as she had in Janet's room. Before Mrs. Johansen and her husband Daniel retired, about an hour later, she checked on the children and all was well.

Baby Janet's crib was next to a window that had a three-quarter length curtain. Behind the curtain was an electrical outlet with nothing plugged into it. The aluminum-wired terminals of the receptacle behind the curtain were passing current through to other receptacles downstream in the daisy-chain circuit. The nightlights and the vaporizers were the only items that were plugged into the circuit's downstream receptacles. The total current drawn by those items was about 10 amps, well within the 15amp current rating for this sort of branch circuit.

At about 12:30 a.m., a neighbor returning home saw flames coming out of the Johansen home. He got to his house and woke up his parents, who called the fire department. At the same time, occupants of a car driving on a cross street also noticed the fire. They did a U-turn and headed down the Johansen's street. They saw the flames at one window, blew the car horn, and went and banged on the door of the house. They heard voices from inside, broke a window, and tried to help the people. They could not get near the window of the room where the fire was.

People at a neighbor's house were watching TV in the den when someone started banging on their front door. They went out and saw the flames coming out of the window of the little girl's room. They ran to a window of the boy's bedroom. Ronnie Larry Harrell, a visitor at the neighbor's house, stated to the fire department that:

> *We ran over, and just as I got there they busted out the window of the little boys room. I raised the frame of the window up and started calling to them—and I could hear him in there crying—saying 'I want my mommy'—but I couldn't see in there. I heard him stumble over in the corner, so I came around on the side and broke this window in his bedroom, and he was right at the window—so I jumped (in) up to my waist and pulled him out the window and brought him over here in the driveway and set him on my car.*

Peggy Johansen was found lying face down across the bed in the back bedroom. Her husband Daniel was found face down on the floor in front of the sofa in the den, with "black stuff" coming out of his mouth, but still breathing. Both survived. Their baby daughter Janet did not. Her body was substantially burned by the fire right there at the crib, which was fueled by the wooden crib parts, the bedding, and the curtain. She had been engulfed in flame. The autopsy report states:

> ***General Description****: The body is that of a 15-month old female whose body has been extensively burned, the full thickness skin being lost over certain bony eminences, especially the skull and the sternum. The features cannot be distinguished, and the hands show a contracture and charring from extreme heat. There is a blackening and charring of the tissues all over the body.*

As in the Hersh fire, the actual fire damage was limited, enabling the

## 4. FATAL FIRE—COLUMBUS, GEORGIA

Columbus Fire Department to determine the point of origin and the cause of the fire with a high degree of certainty. The fire started at the wall outlet, next to the crib. The cause was an overheating failure of one or more of the receptacle's aluminum wired connections, which eventually resulted in a short circuit and sparks that ignited the window curtain.

The Columbus Fire Department was aware of the Federal Government's interest in aluminum wiring problems and notified the CPSC. The agency performed its own investigation that confirmed the Department's conclusion as to the fire's point of origin and cause. A CPSC investigator examined the remains of the receptacle and identified evidence of connection failure. Tests of the curtain material demonstrated that it could be easily ignited by the failure events that occurred at overheating aluminum wired receptacles. The CPSC on-site investigator found three other receptacles in the house—none of which were damaged by the fire—that had evidence of overheating at the wire terminals.

The Johansen house was built and wired with aluminum in 1971. By that time, the various elements of the electrical industry were well aware of the abnormal electrical fire safety risk associated with residential branch circuit aluminum wiring. It was in April of that same year, trying to emphasize the need for aluminum wire with better connectability, that Southwire's Vice President R.J. Schoerner wrote:

> *A genuine crisis exists in our industry, of a magnitude that jeopardizes the entire aluminum conductor market and the life and property of consumers.*

In that letter, Schoerner stated that the situation was a "most serious matter" and that new standards that UL was proposing to solve the connection problems were "barely minimum." His short letter was written to the 50 or so attendees of an urgent meeting that had been convened by UL

Vice President W.A. Farquhar. The attendees were representatives of the aluminum wire and wiring device manufacturers.

The aluminum wiring crisis had been building for more than five years when Schoerner wrote that letter. Shortly after Kaiser introduced aluminum NM cable for residential branch circuit wiring in 1965, other manufacturers followed suit. Aluminum branch circuit NM cable gained market share for new house and mobile home construction because it cost less than copper cable. But the connection problems had not been solved. Local building code and fire protection officials received reports of overheating failures of aluminum conductor connections of all types, and some local jurisdictions restricted or banned its use. They had no stake in the outcome of the market competition between aluminum and copper wiring—they acted in the interest of public safety.

In mid-1966, just a little more than a year after the first aluminum NM cable went on the market, G.W. Pennington, Chief Electrical Inspector of Dayton, Ohio, issued a directive that effectively banned the use of aluminum conductor for all of the sizes commonly used in houses in that city. The directive also prohibited the use of certain types of connectors with aluminum, even if they had the UL AL-CU markings. Mr. Pennington acted after more than two years of investigation, which included meetings with contractors, suppliers, and manufacturers. Based on many factors, including the experience of Dayton Power and Light Company, he had specific concerns about the adequacy of UL's test procedures for connectors for aluminum. Pennington had written to UL asking for test information and data. UL did not provide test data that satisfied his requests.

R.G. Pullen of the National Electrical Manufacturers Association (NEMA) visited Mr. Pennington a few days before the ban was to become effective. NEMA had been alerted and sent Pullen to determine the reasons behind Pennington's action and to persuade him to allow the termination of aluminum conductors in all connectors listed by UL as being suitable for the purpose. The NEMA representative, essentially acting as a lobbyist for

## 4. FATAL FIRE—COLUMBUS, GEORGIA

the industry, had no specific in-depth knowledge of UL's testing procedures and results. Nevertheless, he said that he was quite sure that the factors that Pennington was concerned about were considered by UL in their testing. Pullen told Pennington that UL's test results were considered confidential, and were not sent to anyone other than the manufacturers of the items tested.

Pullen reported to NEMA that the Dayton, Ohio Chief Electrical Inspector's objections were based on concern as to how the aluminum connections would perform in the future. Pennington was certain that, under the conditions of normal use, aluminum connected to some types of UL listed AL-CU terminals available at that time would result in fires in the future.

In his trip report, Pullen noted that another NEMA field representative had reported on a trip to California where the same issue had arisen—failure to accept UL listing for aluminum connection types as assurance of safe long-term performance. Pullen concluded his trip report by stating, "From this information it would appear that the problem is a general one and not limited to the jurisdiction of a single Chief Electrical Inspector."

Failure and fire reports involving aluminum wire and cable connections of all sizes had been coming at an increasing rate for many years, creating concern within the industry. In November 1966, NEMA asked UL to submit a proposal for a "fact finding" field survey study of wire and cable termination failures. It then took more than a year and a half for UL and NEMA to agree on the scope and contract for that study, and yet another year elapsed before the agreed-on survey questionnaires were actually mailed.

Early in 1970, three and a half years after the "fact-finding" study was initiated, a NEMA committee was established to deal with the data that it had developed. The survey returns showed seven times as many aluminum conductor connection failures reported than for copper connections, in spite of the fact that copper conductor was used in the majority of installations. By that time, however, the UL survey was useless. The slow-moving "fact finding" project had been outpaced by growing awareness of the problem, an increasing number of bans on use of aluminum wiring, and some outrage in the public and trade press. A proposed second phase of the UL "fact finding" study was cancelled by NEMA when it became obvious, without any survey, that the industry had a public safety problem, a product liability crisis, and a public relations crisis, all related to the failure and overheating of aluminum building wire connections.

Schorner's letter reflected only one of many voices within the industry arguing for safe performance of aluminum connections. There were proposals within the industry that aluminum building wire products should be voluntarily withdrawn from the marketplace until safe long-term aluminum connection performance could be assured.

The safety concerns and the failure data were not shared with the public. The industry's public face continued to insist that the UL testing was rigorous and that aluminum wire and cable could be used safely. UL, NEMA, and the Aluminum Association became proactive, working together in a coordinated way to overcome local bans and adverse publicity regarding aluminum wiring. Local and State bans that were proposed by homeowners, inspectors or legislators were countered by polished presentations from the industry "experts." J.P. Moran, who had put forth Kaiser's strategy of "taking a calculated risk" by selling aluminum while simultaneously engineering solutions to its connection problems, now was employed by the Aluminum Association. He became a frequent presenter at public meetings and hearings, working against adverse publicity and proposed bans.

To people outside the industry, including home builders and buyers, the industry portrayed aluminum wire as equal in safety to copper wire. Viewed in the most critical light, UL is seen to have been an active partner and protector of the industry in the promotion and marketing of aluminum wiring, even after its abnormal fire hazard risk had been well documented. For this product, at that time and since that time, UL has not been the

# 4. FATAL FIRE—COLUMBUS, GEORGIA

independent watchdog of electrical safety that it portrays itself to be and that the public and the trade believe it to be.

Leviton, the manufacturer of the receptacle at the point of origin of the Johansen fire, was one of the defendants in the resulting product liability lawsuit. When Kaiser first introduced aluminum NM cable in 1965, the wiring device manufacturers were somewhat taken by surprise. Leviton and most of the other wiring device manufacturers received questions from their distributors and field salesmen as to whether the company's switches and receptacles could be used with aluminum wire.

UL provided a quick answer, sending the wiring device manufacturers and anyone else who inquired essentially the same letter that they had supplied to Kaiser in 1958. The letters stated that UL considered aluminum wire and copper wire as being interchangeable on all UL listed wiring devices. There was, in 1965, still no testing by UL to provide a sound foundation for that position. The wiring device manufacturers did not do their own testing to confirm that their products would work safely in long term residential service with aluminum wire. The manufacturers relayed the UL letters to their salesmen and distributors, as proof of UL's position, so their product sales would not suffer in regions where aluminum wiring took hold.

Some electrical inspection departments were resistant. They insisted on following UL's dictum that required AL-CU markings on any equipment used with aluminum conductor. When that came up as an issue, UL simply allowed the wiring device manufacturers to stamp their products with AL-CU or AL/CU marking if they wished. No special testing was required. Electrical inspectors and installers commonly believed that the devices with AL-CU stamped into the mounting strap had passed special tests to earn the marking. It was not true. That deceptive (and perhaps even fraudulent) practice was continued by UL and the manufacturers until mid-1972. M.J. Weitzman, Leviton's chief engineer, wrote in 1973 to a homeowner (who also was an engineer) that the AL-CU marking on receptacles was a "marketing gimmick".

Considerable confusion remains even today as to the AL-CU marking on wiring devices. In 1973, UL published a pamphlet with its final word on installation, including an inconspicuous note that receptacles and switches marked AL-CU should *not* be used with aluminum wire. However, electrical and home inspectors often state that aluminum wire in a home is "properly installed" if the wiring devices have the AL-CU marking. Illustrating the current confusion as to the AL-CU marking on wiring devices is the 2015 edition of the book, "Electrical Wiring—Residential", by R.C. Mullin and P. Simons, which states (p.101) that, for aluminum wire, "Take special care when changing out switches and receptacles or making splices. Use wiring devices that bear the AL/CU marking."

With or without AL-CU markings, most wiring devices manufactured in the U.S. since 1966 have wire terminal screws made of steel. The change from brass to steel for the terminal screws substantially increased the probability of hazardous failures, particularly with aluminum wiring. In 1966, with copper prices high, some manufacturers of receptacles and switches asked UL to allow the use of steel terminal screws instead of brass. Along with the increase in the price of copper had come an increase in the price of brass, which is an alloy composed of copper and zinc. Steel screws would be less expensive. UL agreed to the change, and required that the steel screws be plated with either zinc or cadmium. No testing was performed by either UL or the wiring device manufacturers to determine whether the change had any effect on the long-term performance of the wire terminals.

In 1971, the same year that the Johansen house was built, UL performed its first substantive testing of aluminum-wired receptacle screw connections. The results clearly showed that zinc plated steel terminal screws were the most failure prone with aluminum wire of any materials combination that they tested. In fact, because it represented the worst case, an ordinary Leviton receptacle with zinc-plated steel screws was selected by UL to be used for evaluating the relative connectivity of different aluminum wire

## 4. FATAL FIRE—COLUMBUS, GEORGIA

types. That worst-case combination was installed throughout the Johansen home, including the receptacle at the fire's point of origin.

Well before the Johansen house was built, UL had the justification, the opportunity, and—most people would assume—the responsibility to halt the installation of aluminum branch circuit wiring in houses and mobile homes until the connection safety problems were solved. UL's top executives, President H.B. Whitaker and Vice President/Chief Electrical Engineer W.A. Farquhar were both heavily involved personally in this issue over the years. They called the shots. They consistently acted on behalf of the aluminum wire producers, resisting pressure from the public, from government agencies, from within the electrical industry and from within UL itself to withdraw its listing of aluminum house wiring.

Peggy and Daniel Johansen were interviewed for a segment of the ABC Television show "20-20", which aired in April, 1980. Andy Farquhar appears in the same segment, outside of his home, where he was confronted by investigative reporter Geraldo Rivera. Farquhar declined to be interviewed. Rivera appealed to him on the basis that UL had refused multiple requests by ABC News to provide anyone for an interview, and if he wouldn't speak to them, nobody would be presenting UL's side of the story. Farquhar still refused.

Rivera's confrontation with Farquhar in front of his house was a sort of tabloid reporter tactic, putting Farquhar and UL in a bad light when it aired on the show. The unfavorable image appears to have been well deserved. Whitaker and Farquhar worked hard to maintain the impression that UL was the competent guardian of public safety. To maintain that image, they would not take any corrective action that could be interpreted as admitting that a mistake had been made. They rejected even the simple and direct act of mandating a return to brass terminal screws, which would have substantially reduced the aluminum wiring fire risk in new installations.

By 1971, with more and more burnouts of aluminum-wired wiring devices being reported, the wiring device companies were pressing for a moratorium on the use of residential aluminum wiring. Their own belated testing confirmed the hazardous aluminum termination problems that Kaiser's C.G. Sorflaten had told them about in 1958. At the meeting that prompted Schoerner's letter, in which he stated that the lives and property of consumers were being jeopardized by aluminum wiring, the Leviton representative reportedly stated that nothing that they tried worked properly with aluminum.

# 5.
# "SHOOTOUT" AT TRAVELERS HOTEL

In March of 1971, state-wide restrictions on aluminum wiring were on the verge of being enacted in at least two states, Washington and Colorado. A growing number of local bans had been enacted or proposed across the country. Three wiring device manufacturers had gone on record proposing that UL withdraw its approval of residential aluminum wiring. One of them was Leviton, whose receptacles would soon be wired with aluminum in the Johansen house. Another was Rodale, a relatively small electrical equipment manufacturer, and GE was the third.

E.W. Roberts, GE's attendee at an October 1970 NEMA (National Electrical Manufacturers' Association) meeting, wrote the following about a presentation on aluminum wiring issues surfacing at IAEI (Int'l. Assoc. of Electrical Inspectors) regional meetings:

> *Dick Shaul of the NEMA staff reported on his attendance at every IAEI Annual Sections meeting during the year regarding the action of the electrical inspectors to problems with aluminum wire.*
>
> *The Northwest Section of I.A.E.I. ... expressed the firmest position against aluminum by voting to reject the Code proposal allowing connection of aluminum wire to wiring devices ...*
>
> *... The Southern Section meeting reports that almost all inspectors indicated that they have problems with aluminum connections.*

*... The President of IAEI, Dick Lloyd, says that the problem of aluminum connections is serious and far reaching.*

*U.L. has the message, and Andy Farquhar remarked that the time has long passed when we need more surveys as to whether or not there is a problem with aluminum connections and it is time to do something.*

Half a year later, Farquhar, UL VP and Chief Electrical Engineer, convened a meeting by telegram on extremely short notice. It was sent on April 12, 1971 to companies across the country, announcing a meeting for Thursday, April 15, at the Travelers Hotel near LaGuardia Airport.

The announced subject of the meeting was a proposed change in UL's specification for aluminum conductors to achieve a "more reliable termination of aluminum conductors to wire binding screws." The meeting invitation was sent to all of UL's "listees" (clients) that manufactured wiring devices or aluminum rod and wire. The telegram stated that the meeting was being called "because of the urgent nature of this problem."

Mailed letters, UL's normal way of communicating, were sent on the same day to the same companies. Each letter included a copy of the text of the telegram. The subject line of the invitation letter read "URGENT—MEETING ON SMALL ALUMINUM CONDUCTORS THURSDAY, APRIL 15", all in upper case (which even then was "shouting").

For several years prior to Farquhar's urgent meeting call, Southwire, R.J. Shoerner's company, had been producing a proprietary aluminum alloy that seemed to be more connectable than the more common "EC" (electrical conductor) grade of aluminum. It was marketed under the tradename EEE ("Triple E"). Southwire had been selling Farquhar on the proposition that their EEE alloy held the answer to the poor connectability of aluminum wire.

UL was under mounting pressure from all sides to take substantive

## 5. "SHOOTOUT" AT TRAVELERS HOTEL

action on the aluminum wire fire safety problem. Public safety officials across the country were demanding that UL and the industry either solve the problem or take aluminum wire off the market. The proposed state-wide ban on aluminum wire in Colorado had been put on a temporary hold when industry lobbyists promised prompt action to resolve the problems.

UL had no test results of its own on which to base decisions. Nor did its engineers have the results of research and testing that various manufacturers (other than Southwire) had done or sponsored. Test results were generally treated as confidential or proprietary. Negative results, which are important in the technical decision-making process, were virtually never disclosed to UL. In 1958, when Kaiser asked UL to give a nod of approval for the use of aluminum wire with ordinary wiring devices, the company did not reveal that most devices it tested yielded poor results.

With nothing better to offer, as outside forces were clamoring for immediate and effective action by UL, Farquhar proposed the physical properties of Southwire's EEE as the basis for a new specification for aluminum wire. That did not go down well with the other aluminum companies. Many would be hard pressed to meet the new requirement without using Southwire's proprietary composition and processing to manufacture the wire. Southwire was not about to give it away free on behalf of public safety. Others would have to buy the aluminum rod or wire from Southwire, pay them for a license to produce the same product, or develop their own alloys that would meet the same UL requirements without impinging on Southwire's patents.

The "holy grail" of aluminum wire research—not yet achieved—has been the development of an aluminum wire that could consistently equal copper in connectability without use of abrasion and inhibitor. Toward this goal, most of the extensive research activity by companies and laboratories across the country has centered on the mechanical properties of the wire. The underlying problem, the insulating aluminum oxide, was rarely measured, acknowledged or addressed.

Key researchers in this field, after years of frustration with inconsistent results, came to the conclusion that the only way to achieve reliably good connectability with aluminum wire was to eliminate the insulating oxide. That is exactly what surface abrasion and use of corrosion inhibitor accomplishes. It is an effective but costly installation technique.

Tin-plated, nickel-plated, and copper-clad aluminum wire products were developed to achieve that same objective. But they were not well accepted by the aluminum wire producers. Most manufacturers of aluminum wire wanted to continue producing the poorly-performing EC grade of wire, which they knew how to produce, had the capacity to produce, and would be able to sell at a good profit margin at the lowest price in the marketplace. Aluminum alloy wire would cost more and plated or copper-clad aluminum wire would be even more expensive.

Test results provided by Southwire demonstrated that its EEE alloy was indeed a good performing aluminum conductor, but not equal to copper. One set of test results from Southwire, on receptacle terminals, showed 34% failures with EC aluminum, 3% failures with EEE, and 0% failures with both copper and copper-clad aluminum wire. Significantly, some samples of EEE wire have a conductive surface, in contrast to the insulating oxide on the surface of other alloy and EC wires. Although that was known to be a very important factor, Farquhar's proposed new specifications did not include a surface conductivity requirement.

The short notice for the meeting did not give any of the industry factions time to develop a coordinated strategy and response to the proposal. Wire manufacturers assessed whether they could make a product that would meet the new specifications. Many could not. Kaiser's S.G. Roberts wrote that the new UL specifications spelled trouble for the company, saying, "This reads 'panic'!" Kaiser management resolved to get UL to water down the proposed new requirements to encompass a product it knew that it could produce.

In addition to UL, NEMA, and the Aluminum Association, 21

## 5. "SHOOTOUT" AT TRAVELERS HOTEL

companies were represented at the meeting, mostly aluminum wire and wiring device manufacturers. It was a big and cumbersome meeting, with about 60 attendees spanning a wide range of education and job responsibilities. There were company executives, marketers, engineers, technicians, and lobbyists. This was perhaps the first time that UL and the manufacturers of the wire and wiring devices sat down together in the same room to try to deal with their common problem.

According to W.H. Roemer, an attendee representing Cerro, a wire and cable producer:

> ... Mr. Farquhar opened the meeting by reviewing the history of reported fires in the west coast area which has led to several deaths wherein the Fire Marshalls have assessed the cause to termination failures where circuit size aluminum conductors have been involved.
> 
> As a result of difficulties in termination of aluminum conductors, there is a growing movement toward banning the use of (aluminum) building wire in sizes #6 to #12 AWG in the California, Colorado, Utah, and in some of the New England areas.
> 
> ... It became apparent to us that even if the proposed requirements are met, there is no positive assurance that in actual field use, termination problems would not continue to exist due to improper terminating methods, environmental conditions, misapplication etc. Accordingly, we suggested that in the interest of public safety, UL consider temporarily withdrawing recognition of circuit size aluminum conductors for building wire until such time that sufficient test data could be accumulated to show that aluminum conductors were safe for continued use. No action was taken on our proposal, but Hatfield, American Insulated, Columbia, and the Leviton Wiring Device representative indicated support for our proposal.

B. Hondalus, representing aluminum producer Reynolds, wrote in his meeting report:

> *I went to the meeting expecting it to be another aluminum rod producers' meeting similar to several recently held ... I was most surprised to find 'everyone' in attendance—rod producers, independent wire manufacturers, end users, and wiring device manufacturers.*
>
> *... Such a large and varied meeting called on such short notice was most unfortunate in that there was little semblance of unity among the aluminum producers toward the rest of the attendees. Much 'dirty linen' and 'commercialism' was needlessly aired.*
>
> *Underwriters started the meeting by stating that additional aluminum building wire termination failures had occurred including a number of fires. The state of Colorado almost passed a law prohibiting aluminum building wire in all sizes. The state of Idaho was also very much concerned. Several local codes had outlawed aluminum building wire, etc. Underwriters was under extreme pressure to do something more positive, even withdrawing label approval from aluminum circuit sizes. It was obvious that Underwriters was most fearful of government intervention if they failed to act promptly. Withdrawal of label approval appeared to be the last thing they wished to do if any other action was available to them.*
>
> *... Several strong feelings passed through the assembly much like a tidal wave, but nothing ever came to a vote. Early in the meeting feeling ran strong to adopt the new specification immediately. Shortly before lunch the feeling had changed to temporarily ban the use of aluminum in circuit sizes. After lunch, discussion began on each specific specification with intent being to find acceptable compromises.*
>
> *... it was getting late in the day and Southwire could see that the action being taken was directed at a tighter tolerance EC rather than their proposed new alloy. At this point Roger Schoerner of*

## 5. "SHOOTOUT" AT TRAVELERS HOTEL

*Southwire took the floor and made an impassioned plea for his remaining hope—the thermal soak test. If that, too went down the drain he was lost with EEE. His presentation was most effective. He stated that his test data proved positive that thermal stability was absolutely necessary for good terminations. He offered to pay expenses for everyone to come to Carrollton where he would conduct a seminar to instruct and prove his claims. He pleaded to outlaw aluminum completely rather than adopt a new EC specification. At this critical point, he got strong support from Alcoa who flatly stated that unless UL went to an alloy other than EC the problem would not be solved.*

In his meeting report, Hondalus stated his personal position on several key issues—opinions that he had not expressed at the meeting. Among them was that he would not suggest taking circuit sizes of aluminum wire off the market, "but neither would I object to it."

As for his view of the positions of some of the other companies:

*... Southwire is ... well supplied with supporting laboratory data. Their commercialism is thinly disguised.*

*Kaiser is wishy-washy ... They really seem confused at this point.*

*General Cable objects to everything.*

*Leviton is violently opposed to everything aluminum. They claim nothing will work including EEE.*

*Cerro has ceased production of aluminum building wire purportedly because of the field problems.*

*Columbia ceased production of aluminum building wire but then started again using EEE.*

Additional details, from a meeting report by D.S. Medrick of Anaconda are:

- Alcoa agreed on the severity of the problem.

- Farquhar mentioned that UL did not know if aluminum with the proposed characteristics will improve connection reliability.

- Leviton projected that the improvement with alloy aluminum would be reflected by fires starting to occur in 7-8 years with the newer alloys instead of 2-3 years with EC.

- There was criticism of UL for not withdrawing approval of aluminum circuit size wires when the problem became identifiable.

- UL stated that, if something isn't done, this problem may be taken out of UL's hands. There is need for UL action. Reference was made to Mr. Jensen, Director of the U.S. Dept. of Health, Education, and Welfare, who was quoted as saying that private industry hasn't done an adequate job in providing strong enough standards for the building wire in the field.

No agreement on Mr. Farquhar's proposed new requirements was reached at that meeting, but there was an agreement to meet again.

A few days later, J.W. Mitchell, an executive of Essex, one of the wire producers, wrote to UL's W.A. Farquhar. He protested any change whatsoever to the aluminum wire specifications unless it provided a complete solution. He challenged UL's proposed new set of requirements on the basis that it was scientifically inadequate and commercially inspired by Southwire, writing:

> *Again and again, we come back to the same theme running throughout this situation, U.L. is attempting to change material specifications based on expectations and presumptions in the face of the*

# 5. "SHOOTOUT" AT TRAVELERS HOTEL

*fact that there is no evidence presently available which correlates these material specifications with the connectability of aluminum conductors.*

The public was not told about the aluminum wire connection burnout problems. Most contractors, electricians and electrical inspectors continued to do what they had been doing since Kaiser introduced aluminum house wiring in 1965. The electricians connected aluminum wire to wiring devices using the installation practices that were common in the trade or called for by manufacturers' instructions. The electrical wiring was checked and accepted by inspectors, who looked for conformance to the applicable codes and standards in the general design, for the presence of UL labels on the components, and for a satisfactory level of installation workmanship.

While the meeting was going on, the aluminum connections in the Hersh home and in the Beverly Hills Supper Club deteriorated a tiny bit more, and another day of progress was made toward completion of the Johansen house. For single-family houses, such as the Johansen and Hersh homes, about $15 to $50 per home was saved by the builders who used aluminum wire instead of copper wire.

# 6.
# A FIRE CHIEF'S PERSPECTIVE

Three months before the UL Travelers Hotel meeting, Captain Carl Duncan of the Huntington Beach, California, Fire Department addressed a meeting of regional code officials and electrical inspectors. Duncan served as an onboard electrician in the navy submarine service before joining the fire department in 1964. He had training in basic electrical theory and submarine electrical systems.

Attendees at the January 20, 1971 meeting included some electrical industry representatives. A transcript of Duncan's talk was made from a tape recording and sent to NEMA. Following, from that transcript, are portions of his presentation.

> *... Gentlemen, the termination of aluminum conductors, I'm sure, as you see it, is much different than the way we see it. The slide on the screen represents the way in which I found it the morning after a fire where an individual in our city expired. There's no excuse for that death. ... the cause of the fire was determined to be directly at a convenience outlet termination.*
>
> *... In this particular house, the destruction was almost complete to the interior. We assume approximately a $40,000 fire loss and life loss to one individual. Many people say that the action we have taken in Huntington Beach in prohibiting the use of aluminum wire is a little extreme and some of you relate it to the fact that our*

*assistant building director had an incident in his mobile home which prompted that action. Believe me, that is not the case.*

*We have documented, since last June, some 25 incidents of fire, primarily minor in nature, but many of them with a high degree of destruction.*

*This is a picture of another house where some minor damage occurred, only to about $12,000 ... The burn pattern is the same: the charring of the studs on the interior; one bay affected where the termination to an outlet was made, relatively easy to define; again, total destruction to the outlet, to the box, and to the upper story and attic of this particular home.*

*... People say there is no problem existing in aluminum wiring applications. I have about 1500 to 2000 homes in Huntington Beach and these people would gladly take a stand about that. As you may or may not know, our action in Huntington Beach and in 12 other cities in Orange County has been to prohibit or restrict the application of aluminum conductors in branch circuit applications in single family and duplex occupancies.*

*... The problem exists, gentlemen. We know it exists. It has jumped up and bitten us. We in the fire field are very concerned, not only with fire suppression but with the prevention of fires. I don't think the economic aspects of aluminum wire are worth one life loss, let alone any more. I know there have been more because I have received reports from other communities, other counties, other states.*

*I think it is past due time that we take some action to remedy this problem. In all due fairness to you and to all distinguished manufacturers and inspectors who are here, I realize that there are a lot of other considerations to be made, a lot of technical information available, a lot of tests being conducted, but the fact remains that the method is approved. It is allowable and it is showing some problems. We know from correspondence with UL that they have appointed*

## 6. A FIRE CHIEF'S PERSPECTIVE

*new committees to concern themselves with the approval of this application. We know that the manufacturers are working on new alloys, new materials, and that there are new standards coming out daily. Gentlemen, as we sit here, there are carpenters and electricians out there framing homes and wiring houses with the material that we know by indication and by evidence is not acceptable.*

*Some action should be taken at this time to eliminate the number of hazards which we are building into the new construction phases of our single family, duplex, and mobile home occupancies. We are not going to have any more than 1500 in Huntington Beach with this problem. I am drafting correspondence to every one of these home owners and trailer coach owners who may be exposed to this problem with a recommended solution ... one that we feel to be a relatively safe one. The legal eagles can tell me all they want about sticking my neck out, but I stick my neck out daily. I would much rather do this from a letter standpoint to an individual than going into their homes at night when the structure is on fire. Believe me, I've done enough of that as well.*

*... We have been in contact with numerous other agencies that have the same problem. UL admits it is a state problem. They admit it is a national problem. They also admit that they are trying to do something about it. Well, that's fine, but, in the meantime, we are still building these hazards into our homes.*

Duncan later appeared in 1978 at a Congressional committee hearing on aluminum wiring. Rosalind Hersh had been the first to testify that morning at the same hearing. Duncan attended the meeting as a representative of the California State Fire Chiefs' Association, as well as representing Huntington Beach building safety officials. He read a resolution to the committee that the Fire Chiefs' Association had approved unanimously earlier in the year:

> *We, the California Fire Chiefs Association, have found through our collective experiences in our respective jurisdictions that aluminum wiring used as branch circuits in residential homes presents a potential hazard of fire. We have not found that degree of hazard in similar residential homes that are branch circuited with other conductors.*

Duncan presented the committee with a list of 103 hazardous incidents and fires involving aluminum wiring that the Huntington Beach Fire Department had responded to from mid-1970 to the time of the hearing. He testified that his city had been taken to court over its aluminum wiring ban by the Building Industry Association, and, at considerable legal cost, had defended its right to impose the ban.

> *... we immediately sustained legal action on the part of an electrical contractor who was represented by the Building Industry Association, who decided they would take Huntington Beach to task over this issue, and I am sure felt that if they defeated our ability to impose these kinds of restrictions, the prohibitive ordinance, that they would pretty much have carte blanche with existing installations in other jurisdictions. That case, sir, went from an appeals board decision in the city of Huntington Beach to a landmark decision in the Nation stipulating as to building official's right and in fact his responsibility to act in an area where life and fire loss potential exists, and is made known to him. ...*

Congressman Albert Gore (later Senator, Vice President, and then presidential candidate) posed some questions about a document that Duncan had provided. It was written by a TI (Texas Instruments) manager who attended a UL meeting. TI had developed the copper cladding process and was manufacturing and marketing copper-clad aluminum wire. At one point, Gore asked:

## 6. A FIRE CHIEF'S PERSPECTIVE

*This memorandum written in 1970 ... says, ... that the director of engineering with UL has found himself in the awkward position of having to defend aluminum building wire while witnessing all kinds of burned out devices wired with aluminum conductor. ... Why would Underwriters' Laboratories continue to defend this kind of installation in the face of evidence he apparently had available?*

Duncan responded:

*I have asked several of their staff that question on many occasions, and I never received an adequate answer. I would only be speculating if I tried to propose one to you today, and my best guess would be the economic impact initially, and their credibility, secondarily, would be two high consequence points, and I am in the belief that they in good faith, at least for a period of time, believed that the problem could be solved.*

Gore asked Duncan if aluminum wire was being used for new residential installations. Duncan replied that in his jurisdiction it was not, because of the ban, but it most likely was elsewhere.

*I am sure it is being used, and I am sure that there would be an active attempt on the part of industry to use it successfully, to indicate their technology advances in application, and please believe me, I am not opposed to ... the application of aluminum building wire. I am only opposed to its application when it causes fire ignitions. When they stop the fire ignitions, I will be more than happy to address reconsidering the use of aluminum conductor for building construction.*

Gore then asked, "Do you think that the new technology has stopped, is capable of stopping the fire ignitions or not?" To which Duncan responded:

> ... *I can only tell you, as I told Underwriters' Laboratories ... that we will be more than happy to entertain a repealing of the prohibitive ordinance at such time as you indicate to us that technology has provided you with an answer. I have not heard from them since.*

# 7.
# UNDERWRITERS LABORATORIES

*From the outset, I have had especially in view this subject of protection from fire, and I have succeeded in perfecting a system which, in that respect is not only safer than gas, but is, as I will show to your Board, absolutely secure under all and every condition. ... Your obedient servant, Thomas A. Edison*

EDISON WROTE THAT BRAG IN A LETTER TO THE NEW YORK BOARD OF Fire Underwriters. There were substantive concerns about the risks of fire and electrocution with electricity inside buildings. The Underwriters represented insurance interests, and served as the gatekeeper of public safety at the time that practical electrical applications were first introduced. Contrary to Mr. Edison's claim, electrical safety problems did arise.

Over the years since Edison wrote that letter, a system of codes, standards and inspections evolved that strives to assure the safety of electrical installations. Any wiring device, equipment or method that satisfies the applicable standards, testing and inspections is presumed to be safe for its intended use. For manufactured components of residential electrical systems, development of appropriate standards and testing to those standards was most often done by Underwriters' Laboratories Inc. (UL).

UL traces its origin to the time of the Chicago World's Fair in 1893, when fires broke out at the Palace of Electricity. The insurers hired W.H. Merrill, a young electrical engineer, to investigate. With financial support from the insurance interests, Merrill set up a small laboratory near the fair. The lab became the Underwriter's Electrical Bureau. In 1901, the company moved to a new building outside of Chicago and became Underwriters Laboratories Inc. In about 1916, direct financial support from the National Board of Fire Underwriters ended, as the company was able to support itself from fees paid by manufacturers for product testing and certification services.

Shortly after WWII, when UL's third president Alvah Small wrote that the use of aluminum building wire would likely add to the hazards of life, UL's letterhead proclaimed it to be "for service, not for profit" and that it was sponsored by the National Board of Fire Underwriters. Later, sponsorship was by the American Insurance Association. In 1968, the letterhead slogan changed to "an independent, not-for-profit organization testing for public safety." In 2012, UL became a for-profit corporation.

UL's letterhead slogans were derived in part from its founder's principles, which, it says, are: "Know by test, and state the facts. Testing for Public Safety. Our only function is to serve, not to profit." While UL and its employees may not have enjoyed any profit from the aluminum wire experience, Merrill's other principles appear to have been increasingly ignored as aluminum wire made its 20-year transition from outdoor transmission lines to indoor branch circuit house wiring.

UL's management had an obvious commitment to public safety when Alvah Small was its President. He and UL engineer A.F. Matson had discussed the problem of aluminum connections early in 1946. Matson wrote to Small expressing concern, suggesting tests, and indicating that special connectors might be required for use with aluminum conductor.

On receipt of Matson's memo, Small wrote to M.M. Brandon that there would most likely be problems with aluminum wired binding head screw connections which would present a "definite fire hazard." Brandon's title at UL at the time was "Associate Electrical Engineer." A decade later he became the company's fifth president.

In July 1946, the U.S. Department of Agriculture urgently requested a

## 7. UNDERWRITERS LABORATORIES

statement by UL on the approval of aluminum building wire. The request was prompted by the Rural Electrification Administration, which would not provide financing for building construction projects that used aluminum conductors unless they had UL's approval. The USDA noted a "critical shortage of copper." Brandon responded for UL. He took a cautious approach, stating that UL was awaiting test results and would not authorize the manufacturers to label aluminum building wire as "UL listed" unless the performance was satisfactory.

In addition to demonstrating concern for public safety, UL's management and engineers at that time appear to exhibit what is commonly referred to as a "healthy skepticism" toward the information provided by manufacturers. Brandon questioned aluminum conductor corrosion data that ALCOA had submitted. It did not look right to him, and it was decided that UL should rely on its own tests.

The UL testing program that was conducted at that time was intended to point the way to the development of effective accelerated testing to identify connection types that would not fail in long-term service. The results demonstrated weaknesses of many common connection types when used with aluminum conductor. Connections with copper conductor, which were tested for comparison, showed superior results.

UL published a summary of the test program as Research Bulletin #48, in September 1954. That is the only detailed public disclosure by UL of any investigative test results on aluminum connections. Details of additional aluminum wire connection testing by UL in the 1970s have become known only through the discovery process of lawsuits.

In the years since Small's tenure as President, which ended in 1948, UL management appears to have drifted away from its focus on electrical fire safety and towards a focus on serving the industry. A pivotal example occurred in 1951. M.M. Brandon, not yet UL President but having a level of managerial clout, directed that connection test failures were to be disregarded in considering U.S. Rubber's application for aluminum wire listing.

That might have been logical if the UL testing and listing process could assure an end user that UL listed connectors would work safely with UL labeled aluminum conductor. But that was not the case. It was widely known at the time that using UL listed connectors with aluminum wire did not assure safe long-term operation in actual installations. That was why Kaiser found it necessary to conduct its own connection test and certification program.

UL did not overhaul its qualification testing procedures for aluminum-wired receptacle terminals until 21 years later in 1972, and an additional decade went by before UL issued improved test standards for other types and sizes of connectors for aluminum. It took a groundswell of failures, fires, adverse publicity, public outrage, lawsuits and threatened government action to prod UL and the manufacturers to adopt more stringent test standards. The view of an electrical inspector in 1970, as reported by NEMA field representative W.C. Schwan, was one of many that eventually pushed UL to act:

> ... *Although he was aware of the U/L fact finding study on conductor terminations now under way, he deplored the 15 year lag by the industry in solving the problems of terminating aluminum conductors, which he says were apparent from the beginning.*

UL's slow progress toward addressing the aluminum connection problems appears to have been the result of at least three important factors. First, UL did not have a "war chest" to fund a research and testing program of the magnitude required to understand and resolve this complex technical issue.

Second, the industry expertise that UL had relied on in the past to solve other less challenging problems did not provide effective solutions for this one. The efforts of the manufacturers and the highly-respected research laboratories that some of them engaged to help solve the aluminum connection problem were thwarted by the problem's complexity.

# 7. UNDERWRITERS LABORATORIES

Lastly, UL and its management appear to have fallen increasingly under the influence and control of its clients, the manufacturers. For this product, UL was unwilling and/or incapable of taking independent action on behalf of public safety. For UL to take action that would slow or stop the installation of aluminum wiring in building construction, even when it became clear that it created a unique fire hazard, was out of the question for the two key UL executives who were in command from the mid-1960s to the mid-1970s: President H.B. Whitaker and V.P./Chief Electrical Engineer W.A. Farquhar.

As President and V.P. of UL, Whitaker and Farquhar had the authority required to overcome the difficulties. With regard to funding, they certainly had some say in budgeting UL's financial resources and setting the fees it charged its clients. They could also have reached outside of the involved industry to obtain financing for special projects deemed to be in the public interest. For this problem, UL did not approach government, insurance, or safety organizations for financial help. By depending on the industry to fund special investigations and projects, UL was under industry control in regard to what it could do and how—if at all—important actions and results would be reported to the trade and public.

Investigative testing of aluminum electrical connections was not "rocket science", and the cost was relatively trivial. To start its 1971 study of the connectability of aluminum wire, UL asked the aluminum wire and rod manufacturers for a total of $5,000, to be raised by equal contributions from the companies that chose to participate. In one of the rare UL documents that associates the word "hazard" with aluminum wiring, UL's E.J. Coffey wrote to the manufacturers:

> ... Recently the Laboratories has been advised that the code authority in one county in the state of California has issued an order prohibiting the use of aluminum conductor ... other code authorities are considering similar action.

*In view of this situation and also equally important because of the hazards involved, it is essential that we move promptly to start our testing.*

*We propose to do this work for the aluminum rod producers and divide the cost equally between all companies involved. ... we are not sure that all will be interested. ... However, we wish to confirm that the total cost will not exceed $5,000 and the charges will be applied equally to all companies supporting this effort.*

*Please have an officer of the company ... sign two copies of that form and return them to us with that company's check for $300 as a preliminary deposit. You will note that the cost limit assigned for this form for your share of this investigation is $600 as given in paragraph 4 of the application.*

*Upon receipt of the signed application forms, we will proceed to open an assignment and start the test program.*

To put the maximum cost of $600 for each corporate sponsor into perspective, the average price of a new automobile at that time, 1971, was about $3,500. Start of the test work, clearly recognized as being important for public safety, would have to wait until the sponsors' preliminary deposits came in. UL did not wish to fund the project in advance.

Whitaker and Farquhar could also have bypassed the research roadblock that stymied the industry. If UL's objective was to take action on behalf of public safety, effective steps toward improved performance of aluminum connections were available without the need for further research. UL could have mandated a return to brass terminal screws and the use of inhibitor and abrasion when installing aluminum wire. As an alternate to the use of abrasion, UL could have specified a minimum level of surface conductivity for the aluminum conductor. Or as many inside and outside of industry were calling for, Whitaker and Farquhar could have directed that UL "delist"

# 7. UNDERWRITERS LABORATORIES

aluminum wire—take it off the market—until its problems were solved and its safety could be assured.

Warning the public was another path open to Whitaker and Farquhar. UL's various investigation and follow-up service applications of that time contained a clause advising its subscribers (clients) that it reserved the right to do that, as in this example:

> *It is recognized that, as an independent not-for-profit organization testing for public safety, Underwriters' Laboratories, Inc. will from time-to-time notify the public concerning products which its investigations and tests disclose are extremely dangerous and unsuspectedly hazardous and the Subscriber agrees to the publication of such information relating to products then or previously marketed.*

H.J. Kontje, a former UL Executive Vice President and Chief Engineer, testified in a deposition (on a subject other than aluminum wiring) that he knew of no instance in which UL had ever taken the initiative to advise the public that a hazard existed with a UL listed product.

At the time of the UL meeting at Travelers Hotel, implementing any of these steps unilaterally would have been a major departure from UL's normal way of interacting with its clients. UL did not make changes that tightened requirements without a supporting industry consensus. In fact, a *unanimous* industry consensus was most often required. Through UL's industry advisory committees and the influence of the manufacturers and their industry associations, the details and wording of a UL proposal for change would be massaged and watered down, effective dates would be adjusted and actions might be dropped, according to manufacturers' needs and requests. The so-called "stakeholders" were protected by the UL practices. The public was not.

UL's normal way of interacting with its industry clients was not working well toward solving the public safety problem posed by aluminum wiring.

Nevertheless, Whitaker and Farquhar would not take action to tighten requirements for aluminum wire without industry approval. In contrast, however, they did not insist on industry approval for actions that relaxed or bypassed previously published requirements. Time and again they acted on behalf of Kaiser and other companies without following their usual procedures to gain a consensus.

Kaiser needed UL's cooperation to successfully achieve its marketing objectives. Toward that end, the company's sales engineer R.S. Keith worked toward cultivating a close relationship with UL's Whitaker and Farquhar. Keith apparently came to Kaiser with its acquisition of U.S. Rubber's Bristol plant, and he became the marketing point man for their aluminum building wire products. In late 1957, NEMA's H.M. Dreher received a memo from Keith, to which he responded:

> ... I sincerely hope that you will have occasion to be in New York before too long and, if possible, give me a little advance notice so that I am sure to be here. I will then line up Baron Whitaker for an exclusive get-together—let us say perhaps at the Old King Cole Room where we can pursue the business at hand.
> He is an estimable gentleman—and you should know him.

Roger Keith's glad-handing style was apparently well received. UL engineer T.J. D'Agostino recalled that in the time frame of the meeting at Travelers Hotel in 1971, Keith was among those who could just drop in at Vice President Farquhar's Melville office and chat with him—no appointment needed!

Later, in 1972, when Kaiser's K-140 "alloy" branch circuit wire failed to make the grade in UL's *alloy assessment* test program, Keith and others from Kaiser met with Farquhar. They went away with Farquhar's agreement that UL would change the pass/fail criteria. Farquhar signaled that his intent was

## 7. UNDERWRITERS LABORATORIES

to accept and list at least one submission from each manufacturer. Kaiser's "alloy" was added to the list of UL's approved new alloys.

The wire that Kaiser submitted for the program as K-140 was actually one and the same that had been marketed as KA-FLEX. This EC grade wire had captured the lion's share of the market, and was involved in more than its share of the reported failures and fires. It later was marketed under a new brand name, "KA-FLEX ALR", posing as one of UL's new alloys with superior connectability. But in fact, it had failed the original agreed-upon performance test.

Repeatedly, in the years since Alvah Small retired as President, people who were in high management positions at UL made decisions that overrode the concerns of their own engineers and led them deeper into the aluminum wire problem. One such decision involved Vice President K.S. Geiges and H.B. Whitaker (who became UL President five years later). In 1959, concerns surfaced as to the continued blanket approval of wiring devices for use with aluminum wiring in the absence of any testing. Geiges wrote to Whitaker, who was charged with making a decision:

> *... The recent enthusiasm for use of aluminum wire is confined to the larger sizes and we do not have the problem of binding screw terminals except in special instances, as, for example, in the Kaiser building already mentioned.*
>
> *As a suggestion for the handling of this item I believe that a note [should be] placed in Engineering Data indicating brass binding screw terminals have been considered acceptable with No. 12 and larger aluminum in the past and until such time as there is evidence of a more widespread use of the smaller sizes of aluminum conductor, we plan to take no action with respect to the investigation of such terminals for aluminum wire. This would not contribute anything toward improvement of terminal connections, but it would clarify the present status of wiring devices.*

Whitaker responded:

> ... I believe your suggestion to include some information along this line ... is a good one, and we will proceed in this direction.
>
> ... We share your view that the use of aluminum building wire is still more theoretical than actual, and accordingly we feel that the concern of our engineers in the wiring device field anyway is primarily academic.

Both Geiges and Whitaker recognized that testing was required, but concluded that it could wait until there was evidence of increased use of aluminum branch circuit wiring. But they did not follow through when Kaiser introduced branch circuit aluminum wire into the home construction market in 1965, six years later. By the end of 1969, with more than a million new homes and mobile homes wired with aluminum, UL still had not done any qualification testing of wiring devices with aluminum wire.

T.J. D'Agostino was the project leader for the testing that UL finally undertook in 1971. He recalled attending a meeting in 1979 with former UL President H.B. Whitaker and others to prepare for a deposition in the Beverly Hills Supper Club fire litigation. D'Agostino was handed a letter that he had written to Whitaker in about 1973. In it, he recommended taking aluminum branch circuit wire off the market in the interest of public safety. After reading the letter to refresh his memory, D'Agostino held it up for Whitaker to see and asked him, "Baron, how about you, do you recall the letter? I never received a reply." Whitaker took the letter, read it and quietly said, "Yes, I recall the letter and I, as President, saw fit to not reply. Basically, I crumpled it up and threw it into the waste-paper basket."

# 8.
# MOBILE HOME FIRES

Victims: Candy Belden, 9, and Shelly Belden, 16
*(Belden Family photos salvaged from fire remains)*

In August 1979, the Wittmann, Arizona mobile home of Janet and Irwin Belden was destroyed by fire. Their two daughters perished. Local and Federal investigators determined that the fire was caused by an aluminum wire connection failure.

In an interview aired on ABC's *20/20*, April 3, 1980, Mrs. Belden stated:

> *I went into my girls' room and kissed them both goodbye and left for work. At ten minutes to six I received a call ... that the trailer was on fire. I came home. ... the trailer was gone, my girls were gone, and Erv was laying out on the road. And when I left at five o'clock everything was beautiful ...*
>
> *I know that there are a lot of tragedies that have happened with aluminum wiring, and I also understand that the industry knew*

*that it was unsafe, so they shouldn't be sold, they should never have been on the market ...*

Mobile homes account for about 6% of the housing units in the US. There are more than 8 million of them, housing an estimated 20 million people. Most mobile homes built between the mid-1960s and early-1970s were wired with aluminum. The manufacturers saved an estimated $9 to $17 per home by using aluminum wire instead of copper.

In 1966, the year after aluminum NM cable appeared on the market, mobile home manufacturer Hensley had so many problems with aluminum wire that it returned its remaining stock for refund and switched back to copper. Many of the problems had to do with poor quality control of the aluminum NM cable. Some of the wire was too hard and brittle, and there were defects in the wire insulation and the cable jackets. The defective cable came from several manufacturers.

In March of 1969, A.G. VandeWiele, owner of a mobile home, wrote to the Chief Deputy Fire Marshall of Washington State describing hazardous incidents at aluminum-wired receptacles:

*A possible serious fire hazard has come to my attention and I want to pass this information on to you in hopes it may be of value.*

*... a very unpleasant odor ... would come and go. ... in a few days was able to determine it was caused by heat in one of the plug-in receptacles. ... the only time the heating and odor occurred was when the electric iron was being used in another plug-in about ten feet further down the wall.*

*... I noticed that ... it was so arranged that the electricity had to pass into and back out of the receptacle in order to go on to the next receptacle where the iron was plugged in. About two months later the same problem occurred on one of the plug-ins in the bedroom*

*... one of my neighbors had a fire, which he explains was caused for*

## 8. MOBILE HOME FIRES

*the same reasons ... After his fire, while carpenters were repairing his home, they again experienced the odor and heating in another receptacle in the house.*

*I am wondering how many fires have occurred due to this type of condition and also wonder how many more will occur before corrective measures have been taken.*

Mr. VandeWiele may have been especially sensitive to the possibility that there might be a wide-reaching fire safety problem. He worked in the State's Insurance Commissioner's office. His letter was relayed to the State's mobile home inspection department and a copy was sent to E.H. Bastedo, Kaiser's regional sales manager. Mobile home manufacturers accounted for the major portion of the KA-FLEX sales in his territory. Bastedo wrote to others in his company:

*The attached letters are frightening. You will recall that letter from Montana. I have heard of trouble in Alaska. Then a couple of weeks ago the City inspector at Hoquiam, Washington told me of one fire and one case of hot receptacles in that town.*

*Maybe this is not our problem, but I believe it is now serious enough for a full-scale investigation.*

With ever more complaints and fire incidents reported, the State of Washington Inspection Office issued Policy Letter No. 8, November 27, 1970, which required that receptacles in mobile homes be "pigtailed". That practice uses splicing connectors to attach short copper wire "pigtails" to the aluminum wires. Then the wiring devices are connected to copper wire instead of aluminum wire. Whether that aluminum wire termination method is safe depends on the performance of the splicing connectors that are used. Many of the available splicing connectors that UL listed for pigtailing with aluminum wire were even more failure-prone than the screw terminals. This was not widely known at the time. UL had not tested them for that application.

Installing receptacles with copper pigtails took more time than connecting the aluminum wire directly to the terminals on the device. The mobile home manufacturers opposed the pigtailing requirement for aluminum wiring because it would add to their labor cost. Their trade association and a regional mobile home distributor requested a hearing on the matter. It was convened on January 16, 1971. Attending the hearing were mobile home manufacturers, code enforcement officials from several states and representatives of UL and the aluminum wire manufacturers. Kaiser's R.S. Keith wrote in his trip report:

> *... Similar requirements are or will go into effect in Oregon, Idaho, and Utah ...*
>
> *... John Hewett opened the hearing with a frank explanation of the reason for the pigtailing requirement, namely, a continuing spread of arcing and burning of aluminum conductor connections at 15 and 20 amp wiring devices and that action had to be taken. Much pressure was coming from the fire marshals, mobile home owners and others which could cause action more drastic than the proposed pigtailing unless the industry had something to offer now. ...*

The Inspection Office was not swayed by the protests of the mobile home and electrical industry attendees who argued against the pigtailing requirement but did not have a workable alternative. The restriction remained and Washington State inspectors started enforcing it.

Two months later, J.P. Moran, who had put forth Kaiser's "calculated risk" strategy, met with W.E. Dell, Supervisor of the Washington State mobile home inspectors. This was one of Moran's first missions in his new job as an employee of the Aluminum Association. His trip report states:

# 8. MOBILE HOME FIRES

> *They have had a considerable number of fires in Mobile Homes and they are convinced that Aluminum cannot be safely terminated to screw terminals, on the devices which are on the market, especially if such terminations are to be made by the class of help to be found in the mobile home factory. These are not electricians. At the crux of this problem, they feel, is the basic philosophy of MH building, namely you build it in the lowest cost labor market you can find and then haul it to the highest price home market you can find.*

The mobile home electrical fire problem could not have been solved by using higher skilled installers. Even UL employees, definitely higher paid and trained than the mobile home installers, could not consistently make aluminum wire connections that did not overheat. O.G. Wedekind, Associate Managing Engineer from UL's Santa Clara, California facility, had represented UL at the January meeting. About eight months earlier he took the initiative of buying some receptacles locally and installing them with aluminum wire in a wall, just as electrical contractors and mobile home manufacturers would do. As obvious as it might seem, nobody in UL had yet done this. A UL technician (or Wedekind himself) wired the receptacles, tightened the terminal screws, pushed the receptacles into the outlet boxes, fastened them and attached cover plates. It was done according to the accepted practice at the time. Many of the connections overheated immediately at the start of his test.

Wedekind's simple project most likely cost UL far less than his trip expenses to the Seattle meeting. He undertook it in an attempt to personally understand what was causing the rash of reported aluminum wire burnouts and fires. Wedekind's test receptacles were wired with the larger branch circuit size aluminum wire, #10 AWG, which is the size specified for 20 amp circuits. A majority of the burned receptacle reports that he had received involved that same size wire.

Wedekind's test involved only 10 receptacles, installed in the typical daisy chain fashion. His test current was only 15 amps, less than the circuit rating. Current flowed through all four of the aluminum-wired screw terminals on each receptacle.

Wedekind noted several key factors that caused his aluminum-wired receptacles to overheat, among which were:

- The connections loosened when the receptacles were pushed into the outlet box.

- The terminal screws were too small and there was too little clearance for the wire. This, he observed, "makes it difficult for the electrician to make an effective connection … since these terminals were never really designed for this size of wire."

- Many receptacle designs had physical features that prevented the wire from being pressed flat against the brass terminal plate when the screw was tightened.

The overheating receptacles in Wedekind's test were then pulled out of the outlet boxes far enough to be retightened and reinstalled. He wrote:

> *Before being reinstalled, however, we folded each conductor over the back of the receptacle … in such a way that the force exerted by the bending of the conductor during installation could not cause any turning effect on the terminal screw.*

Having retightened the terminals and reinstalled the receptacles using a novel technique, Wedekind then resumed the test. The setup ran automatically, applying current on a cycle of one hour on, one hour off. In spite of all the precautions and retightening, the long-term data from his test shows that additional instances of overheating developed a few years later.

Wedekind's long-term test on these receptacles fell far short of representing actual use. The temperature and humidity range that a receptacle is

## 8. MOBILE HOME FIRES

exposed to in the perimeter wall of a house or mobile home is far greater than inside a room at UL's Santa Clara facility. Additionally, there was none of the normal household activity of inserting and removing plugs to connect ordinary household electrical items. Inserting and removing plugs imparts a mechanical disturbance on the wire connection, since the plug blade contacts are part of the same brass piece that forms the wire terminals. High humidity, wide temperature swings and plug-in/out operations have all been shown to hasten the failure of aluminum wired receptacle terminals. They are the facts of life for a residential branch circuit receptacle. None of these deteriorating factors were present in Wedekind's test.

Both UL President H.B. Whitaker and V.P. W.A. Farquhar later referred to Wedekind's test many times in meetings and hearings, using it to portray the image that UL had thoroughly tested aluminum wiring under actual residential conditions and that the results were good. They never revealed that some of the receptacles overheated when first installed, that additional overheating occurred two years later, that several "wire nut" splices in the aluminum wire test system had burned up, that a total of only 10 receptacles were tested and that the test conditions were far milder than would be experienced by devices and splicing connectors in a house or mobile home.

Wedekind had not hidden the information from others within UL. In May, 1971 he had written to Farquhar, and copied several others, that:

*Attached is a tabulation of temperature readings taken at various times since early last August. Actually, testing was begun early in July, but we found immediate overheating problems with a number of the receptacles and analyzed these for methods of correction before we were satisfied to put the complete assembly under final test.*

Taking corrective action when overheating occurred, Wedekind glossed over the fire safety problem. Had he continued the tests without correcting the overheating receptacles, some of them might have ignited a fire in his test wall. He had come face to face with the problem that people were encountering in houses and mobile homes across the country, but then looked the other way.

In May 1972, Kaiser's C.R. St.John relayed information from their customer Melody Mobile Homes to A.S. Hutchcraft, who later became President of Kaiser. He wrote that the customer:

> ... *experienced hot receptacles and burned-out receptacles in 500 to 600 of 1100 homes they built with wiring devices which had steel screws. ... Melody Homes is switching to copper clad or copper wire ... Mr. Kelly feels U/L has done a terrible job and is blaming aluminum wire when the only problem is steel screws in the wiring devices.*

Later that year, Foremost Insurance Company stopped writing policies for mobile homes with aluminum wiring. Their action was based in part on a study of fire statistics. They noted an increase of mobile home fires that coincided with the increased use of aluminum wiring. American Plan, another insurance company, followed suit shortly thereafter.

By the end of 1972, the use of aluminum wire in mobile homes was in a steep decline. Bans or restrictions on its use in buildings and mobile homes were in place around the country, including all or parts of the states of Florida, Texas, Maryland, Georgia, Louisiana, Virginia, Tennessee, and New York. Where bans or restrictions did not exist, some builders and mobile home manufacturers were switching to more expensive copper wire or copper-clad aluminum wire on their own initiative. The bans and restrictions applied to new construction. They did not address the fire risk for families already living in aluminum wired homes.

In 1973, a bill was introduced into the U.S. Congress that sought to improve the safety of new mobile homes. The Florida Congressman who had introduced the proposed legislation, L. Frey, Jr., noted that the electrical

## 8. MOBILE HOME FIRES

and mobile home industries defended questionable practices by stating that they met the applicable standards. He specifically mentioned aluminum wiring as contributing to mobile home fires. Congressman Frey pointed out that the standards reflected what was acceptable to the industry stakeholders, and not necessarily what was safe for the occupants.

Also in 1973, only about 35 miles from the site of the 1979 Belden tragedy, a bill was introduced in the Arizona State Senate prohibiting the installation of aluminum wire. There was a committee hearing on the bill on Feb. 4. Representatives of six aluminum wire manufacturers and the manufacturer of copper-clad aluminum wire were at the meeting. J.P. Moran and an attorney represented the Aluminum Association. Once again, O.G. Wedekind came from Santa Clara to represent UL and argue against imposing the ban.

A local professional engineer, F.C. Jones was there to testify for either UL or the Aluminum Association. He had been engaged by UL's VP W.A. Farquhar to testify in case Wedekind could not get to the meeting. Jones was prepared to switch and testify on behalf of the Aluminum Association if Wedekind appeared. He brought with him two prepared statements, one to use for UL and another that he would be able to use if he ended up representing the aluminum manufacturers' trade association. It did not matter to Jones, since UL and the Aluminum Association were on the same team.

The UL and industry people, together with Mr. Jones, met the evening before the hearing to coordinate their proposed remarks. They met again the morning of the hearing and reviewed a series of slides that presented information on how the manufacturers had addressed the aluminum connection problems. Wedekind, Jones and the industry representatives met again after the hearing to discuss follow-up actions.

Arizona did not have its own state-wide electrical code. The UL and industry representatives proposed that instead of banning the use of aluminum wire, the bill should be revised to reference the National Electrical Code (NEC) as being the minimum acceptable standard for new electrical installations in the State. Asked by one State Senator about how the NEC deals with aluminum connections, Wedekind reported that he answered:

> *I also referred to the fact that the general rules relating to the making of electrical connections are covered by (NEC) Section 110-14 and that it specifically treats the special problems associated with aluminum conductor connections.*

The 1971 edition of the NEC was the latest at the time of the hearing. Section 110-14 of that edition—and subsequent editions—is brief. The following statements reflect its total content relevant to branch circuit aluminum wiring connections of the types that were causing fires:

> *Because of different characteristics of copper and aluminum, devices such as pressure terminals or pressure splicing connectors and soldering lugs shall be suitable for the material of the conductor and shall be properly installed and used.*
>
> *Connection of conductors to terminal parts shall insure a thoroughly good connection without damaging the conductors ... No. 8 and smaller solid conductors may be connected by means of wire-binding screws ... or the equivalent.*
>
> *Conductors shall be spliced or joined with splicing devices suitable for the use ...*

Underlying these NEC "motherhood and apple pie" statements is faith in the process by which an independent testing laboratory—UL in this case—determines whether a terminal or connector is "suitable" for use with aluminum wire. Mr. Wedekind's own test experience provided ample proof that adherence to the NEC did not assure safety. His ten test receptacles, all listed and approved by UL for use with aluminum wire, were installed

## 8. MOBILE HOME FIRES

at UL in accordance with the NEC and several of them overheated right from the start at only 75% of rated current.

The only person urging adoption of the ban at this hearing session was a homeowner who had experienced a fire caused by a failing aluminum wire connection. A representative of the Arizona State Fire Marshall's Office, who was called for at Senator Baldwin's request, came in cold on the subject. He had heard reports of fires attributed to aluminum wiring but had not personally investigated any of them.

At a previous hearing on the bill earlier that month, E.R. Thorne, director of the Arizona State Division of Building Codes, had supported the ban, citing aluminum wiring as the cause of numerous mobile home fires. At the same session, H. Blake, field supervisor for electrical inspection for Phoenix, testified about numerous complaints, three cases of injury, and one death due to fire from failure of aluminum wire connections. Senator Baldwin, who had originated the proposed bill, noted that Mr. Blake's job had been threatened because of his testimony at the previous hearing.

The industry prevailed. Kaiser's representative at the hearing, R.L. Shelton wrote, "… we were most successful in amending Bill 1113 to reflect our own interests …." The original ban clause in the bill under consideration was amended to read, as grounds for suspension or revocation of license:

> *Knowingly installing or causing the installation of electrical wiring in any building or structure in this State which does not meet or exceed the requirements of the National Electrical Code of the National Fire Protection Association in effect at the time of such installation.*

By voting for this change, the State Senate committee members expressed confidence in the industry and in the quality of UL's work. There had been some problems, but the industry claimed that new UL requirements were expected to eliminate the aluminum wiring hazard. When the industry representatives were holding the NEC up to the Arizona Senators as if it were The Bible, they all knew that the existing UL certifications for "suitability" of terminals and connectors for aluminum wire could not be relied on. Most were also aware that there was no confidence within UL itself or the industry that implementing UL's new requirements would solve the problem. The NEC's words could only assure safe application of aluminum conductors if the testing authority, UL in this instance, acted effectively and responsibly in the interest of electrical fire safety.

Families were living at risk. Aluminum wire connections in their homes were running abnormally hot, just as Wedekind's simple test had demonstrated at UL's Santa Clara Laboratory. Wedekind took corrective action before continuing his testing. The industry and UL did not provide any corrective action advice to the public, nor did they warn of the hazard. Without corrective action, the hot-running aluminum connections in houses and mobile homes progressively deteriorated. In the worst cases, the process proceeded to fire ignition.

At about 8:00 on the morning of April 16, 1979, Bruce Young returned home from his night-shift job and went to bed. Young Brucie, three years old, also went to bed for a nap. Mrs. Young left for her job, and their 5-year old daughter was at school. The Young's aluminum-wired mobile home, 15' wide by 70' long, had been manufactured by Pacemaker in 1969. They moved into the home, in Concord, New Hampshire, in 1974. Subsequently, they experienced some incidents of overheating receptacle connections. An electrician had replaced the particular wiring devices involved in those incidents. Nothing more.

At about 9:00, Mr. Young was awakened by the sound of the smoke detector. According to the CPSC's investigation report:

> *Realizing that there was smoke in the bedroom, he jumped out of bed. He could hear Brucie yelling "fire." Mr. Young then ran the length of the mobile home into the kitchen where he saw flames ... The*

## 8. MOBILE HOME FIRES

*flames were in the shelf area between the range and the refrigerator ... Running back to his bedroom he attempted to phone for help. He dialed "0" (for the operator) but did not get a response. Again he dialed "0" and at this time heard a "whoosh" sound, and quickly the smoke banked down to chest height. Leaving the phone, he began calling for Brucie, but he was nowhere to be found. He then thought to look under the bed. This is where he found his son who was hiding from the smoke. Mr. Young grabbed his son, and they both escaped ...*

Local and CPSC investigators concluded that the fire started at the aluminum-wired receptacle that supplied power to the refrigerator. The Youngs were very lucky. Nobody was hurt. It was a very close call.

One year later, the ABC Television show *20/20* aired its segment on aluminum wiring titled "Hot Wire." That program sparked a different kind of fire.

# 9.
# TRIAL BY TELEVISION

The "Hot Wire" segment of ABC's weekly TV show *20/20* on April 3, 1980 was introduced by host Hugh Downs saying:

> *Home, to almost everyone, means safety, security, a place to nurture our families. The last thing we want to think about is some danger hidden in our walls, ready to destroy our homes tonight or next year. But for some, it's there—a dangerous kind of wiring. Geraldo Rivera has a special report.*

Geraldo Rivera, an attorney who was an investigative reporter for *20/20* at the time, continued:

> *We're talking about aluminum wiring. Despite five years of bad news and despite the fact that the Consumer Product Safety Commission made it their number one priority, it's still installed in two million American homes and house trailers. This is a report about how and why, and, most importantly, about who is responsible for putting aluminum wiring on the market in the first place.*

The 15-minute long segment was the third one in the program. It followed a piece on anorexia and another on the future of the entertainment business after the imminent introduction of the video disc. "Hot Wire" started off with the Beverly Hills Supper Club fire and litigation, reporting that nine of the companies, including UL and Kaiser, had settled before the case went to the jury. The jury verdict—that an aluminum wire connection failure was not the cause—was not what the companies expected. The narrator states that it was "certainly a major victory for the aluminum industry."

The program interweaved interviews, industry documents, views of burned-out aluminum wired receptacles, photos and videos taken at fire scenes, a demonstration of how a fire starts from a receptacle with an overheating wire connection and more. Connection problems with the larger sizes of aluminum cable were illustrated by mention of failures at the World Trade Center, less than 10 years after it was built. Rivera stated that:

> *ABC News has learned that the World Trade Center in New York City has also experienced serious problems with aluminum wiring. ... aluminum wire terminations of the large cables in the building have been replaced. The job took over a year to complete and cost hundreds of thousands of dollars.*

In the entire segment, only one person interviewed defended, somewhat, the use of aluminum wiring. J. Ross, with the National Fire Protection Association (NFPA), the organization that publishes the National Electrical Code (NEC), was asked if he thought that a mistake had been made by essentially equating aluminum wire with copper wire in the code prior to 1973. He answered:

> *I have got to say that I have been working with and reading the National Electric Code since 1941, and "mistake"? I would never use the word mistake. ... because I don't feel that there is a mistake there.*

Rivera reported that nobody from the industry or UL would talk with

## 9. TRIAL BY TELEVISION 93

them to give the industry's side of the story. His attempt to interview UL's W.A. Farquhar at his home was included in the program, with Farquhar refusing to be interviewed and saying nothing about aluminum wiring. Rivera noted that:

*We asked Kaiser Aluminum at least seven times to make someone available for an interview. They also refused, although Kaiser did say they would make a televised statement five minutes in length if 20/20 agreed both to air the entire statement unedited and then not ask the person making the statement any questions. That, of course, would have been a violation of ABC News policy, not to mention journalistic ethics. Kaiser did send us two letters reciting their position on this issue and in the interest of fairness I'll summarize the company's main points*

*Kaiser maintains that aluminum wiring is safe if it is installed properly. Now, with respect to the Kaiser representative's statement that the National Electric Code is "a political instrument," Kaiser now claims that the word "political" was not intended to suggest that there were any compromises as to your safety. Kaiser, it also should be noted, no longer manufactures this type of aluminum wiring.*

*Now, we've carefully considered Kaiser's position on this issue, but we stand by our story absolutely. If your house or mobile home is wired with aluminum, we suggest you call the Consumer Product Safety Commission's hotline. Their number is ...*

On the *20/20* show one week later, Hugh Downs reported that 55,000 people tried to call the CPSC's toll-free hotline number in the first hour after the segment aired. The CPSC hotline was overwhelmed with incoming calls for at least a week. The 1-800 toll free network on the east coast was bogged down for several days because of the number of callers trying to get through to the CPSC. The *20/20* presentation, broadcast nationally, had been very effective toward publicizing the aluminum wiring hazard.

Kaiser was portrayed by *20/20* as the villain, ignoring its own test results and knowingly marketing a hazardous product. The company reacted by threatening legal action unless the network provided it with an opportunity to rebut the allegations. Kaiser insisted that ABC News broadcast on *20/20* a ten-minute rebuttal that the company would provide. ABC News refused. It responded that Kaiser, UL, and others in the industry repeatedly declined the opportunity to tell their side of the story on the program.

A full-page Kaiser advertisement headlined "Trial by Television" appeared in major newspapers. The ad attacked some practices of TV broadcast journalism, defended Kaiser and aluminum wiring, and called for public action to defend the principles of fairness to the accused on which our country was founded. It read:

> *The American system of justice is founded on a simple principle: The accused has the right to be fairly heard in his own defense, and to confront and cross examine his accuser. ...*
>
> *Yet today, here in America, charges are aired before tens of millions of people without fair opportunity for the accused to respond. They call it "investigative" television journalism. We call it "Trial by Television".*
>
> *Much of investigative television journalism is solid and responsible reporting—but much is not. Many producers of "news magazine" programs too frequently select story segments with their minds already made up about the points they want to make. Then they proceed to select the facts and quotes which support their case. "Interview" opportunities are sometimes provided the "accused." But the edited "interview" format puts the producer (i.e. the accuser) in full control of deciding what portions, and how much of, the*

## 9. TRIAL BY TELEVISION

> *accused's defense the public will be allowed to see. Rarely does this result in balanced and objective coverage.*
>
> *The television production team becomes the accuser, judge, and jury, with no real recourse for the accused to get a fair hearing in the court of public opinion. Yet the viewing public is led to believe that the coverage is balanced and objective. This is a deceptive and very dangerous practice.*
>
> *"Trial by Television", like the kangaroo courts and star chambers of old, needs to be examined. If we decide, as a society, that we are going to try issues, individuals, and institutions on television, then some way must be found to introduce fairness and balance.*

The concept of fairness expressed by Kaiser in this public advertisement stands in sharp contrast to the company's statements and actions outside the public view, in its relentless drive to promote the sale of aluminum wire. For instance, Kaiser's R.S. Keith encouraged the company's salesmen to lobby actively for passage of proposed electrical code changes that would result in more sales and profit for the company. In 1967 he wrote to the sales force that:

> *In effect we are all Technical Marketeers selling Code changes in order to play in our own ball park, or on our own golf course, with our own rules ...*

Kaiser essentially accused ABC of being dedicated to presenting only one side of the story to the public, which is exactly what Kaiser, the industry, and UL had been doing for more than a quarter of a century in the promotion of aluminum building wire. It could be argued that their bias is acceptable because the trade and the public at large expect manufacturers to skew information that they release in favor of their products and withhold anything that is unfavorable.

The Kaiser ad went on to deny the conclusions of the *20/20* program, and it described the actions the company was taking in response. The ad stated:

> *... On its 20/20 segment ... the announcer accused aluminum house wiring of being unsafe, and Kaiser Aluminum of intentionally marketing an unsafe product. These accusations are blatantly wrong.*
>
> *Although we were offered an opportunity to be "interviewed," 20/20 reserved the privilege of editing any part of our statement. Any defense we might have made would be subject to their sophisticated editing techniques, and to their commentary. Since it was evident to us that the producers had already formed their opinions, we declined their offer. How can a defense be fair if it is subject to censorship by the accuser?*
>
> *... We will not allow ourselves to be maligned or misrepresented by any group—even television.*
>
> *Here is what we are doing: 1. We have demanded a satisfactory retraction from ABC TV. 2. We are asking the Federal Communications Commission under their "Personal Attack" doctrine, to order ABC TV to provide us with time and facilities to present our side of the story to the same-size audience in a prime time segment. 3. We have asked Congressman. Lionel Van Deerlin (D. California), Chairman of the House Subcommittee on Communications, to consider Congressional hearings to examine the implications of this increasingly insidious and dangerous practice.*

The ad concluded with an appeal for citizens who believed in the rights of the accused, who believed in the prevention of tyranny by a system of checks and balances, who believed in the rights of the public to get balanced and objective information, and who believed in the American system of justice to write to their elected officials or to Kaiser.

## 9. TRIAL BY TELEVISION

The ABC *20/20* program was not the only one to broadcast a piece on aluminum wiring at that time. WRC-TV in the Washington, D.C., area aired a similar investigative piece that it had produced earlier that same year. The program—*Byline: Lea Thompson*—identified aluminum wiring as a hazard and discussed overheating aluminum wire connections, fires and bans. It talked about the self-regulating industry that was responsible for the product and its lack of checks and balances. The Chairman of the CPSC, Susan King, was interviewed and discussed the objectives of the agency's action against the wire and device manufacturers—to compel the companies to warn the public and correct the hazard. As with the *20/20* investigation, the industry refused to participate. "We've tried to get the aluminum industry's side of the story, but everyone we contacted refused to be interviewed," Thompson stated on the program.

Unlike the *20/20* segment, *Byline* did not focus on Kaiser. The only reference to the company was a quote from a Kaiser document describing its 1964 marketing plan:

*Every day someone brings out a new product, and most of them, whether they finally end being judged good or bad, are time tested at the expense of builders and home owners.*

The piece closed with the following remarks by Thompson and G. Wooley, a homeowner who had experienced several aluminum connection burnouts:

Thompson: *It would be nice to think that this type of thing won't happen again. But the fact is the electrical industry is basically self-regulating, and that what it deems safe the public has to live with. Without a check and balance system, a new unknown system was thrust onto the market, and now we, as the consumers, have to pay the price.*

Wooley: *The price people are going to have to pay ... some of them are going to die, some of them are going to get hurt, from fires, some of them are going to lose their homes, some are going to lose valuable possessions that are irreplaceable, ... the price is fire.*

The *Byline* program was definitely what Kaiser people would call "anti-aluminum." (Kaiser people never used the term "pro-safety".) As with the *20/20* show, the company responded by threatening legal action unless the station aired a retraction and a Kaiser rebuttal. The company claimed that the way the quote from its marketing plan was used implied that the product had not been adequately tested. Kaiser insisted that its aluminum house wiring had been extensively tested before it was sold for home construction.

WRC-TV management rejected the demand and proposed to create and air a follow-up piece on *Byline* that would present Kaiser's position, provided that the company would give them the technical reports of its tests. The station's management was aware that the Kaiser tests illuminated the problems, not the safety, of aluminum house wiring. Kaiser rejected the WRC-TV proposal and insisted on providing its own rebuttal for the station to air. Station management held firm, and that is how the situation ended for them.

Aluminum wiring safety issues were also in the news in Canada. The Canadian public television (CBC) affiliate in Toronto produced a program segment not unlike those that aired in the USA. Subsequently, ALCAN, the dominant Canadian aluminum wire producer and CSA (Canadian Standards Association), the Country's counterpart of UL, threatened to sue the network. They claimed that the problems discussed and demonstrated in the program were unique to the US brands of aluminum wire and connecting devices. They cited a demonstration performed at a laboratory in New York that showed fire ignition resulting from an overheating aluminum wire twist-on connector splice in an electric baseboard heater.

## 9. TRIAL BY TELEVISION

The laboratory and the program's producer had ample proof that Canadian aluminum wire, baseboard heaters and connectors were used for the program's discussion and demonstration. As with the WRC-TV *Byline* segment, CBC management held firm as to the accuracy of their program, and that confrontation ended.

Management at ABC-TV, the parent company of ABC News, did not hold firm. A different outcome developed that sent seismic ripples through the ranks of investigative TV reporters. Geraldo Rivera's closing statement in the *20/20* show, "… we stand by our story absolutely …," was based on an extensive seven-month investigation. The information that was developed included knowledge of documents submitted as potential exhibits in the Beverly Hills Supper Club litigation. They were contained in about 100 loose-leaf binders lined up along the railing near the front of the courtroom.

The arrangement was created for the convenience of the attorneys, but anyone could walk up and thumb through them when the room was open and court was not in session. The producer of the *20/20* "Hot Wire" segment spent time as an observer in the courtroom, looking at the documents and speaking with people involved in the aluminum wiring investigations. Lacking the same in-depth knowledge base, management and attorneys of the ABC News parent company did not share their subsidiary's level of confidence.

After "Hot Wire" aired, Kaiser requested a "reasonable opportunity to respond" that it claimed was guaranteed by the "personal attack" rule of the Federal Communication Commission (FCC). ABC-TV, overruling the objections of its ABC News subsidiary, agreed to provide air time for an unedited Kaiser rebuttal. That was an unprecedented concession in the network television world.

Kaiser was not the first company to have been subjected to what its management considered to be a "hatchet job" by a TV journalist. Illinois Power, Mobile Oil, Hooker Chemical and the American Association of Railroads were among those that had taken issue with specific network TV programs. They had responded in various ways, but were never able to access the same audience and achieve the same clout as the original offending program. Kaiser had scored a breakthrough at ABC-TV in that regard.

In a piece titled "Kaiser Case May Hamper TV News Probers", Associated Press writer Peter Boyer wrote:

> *The capitulation, advised by network lawyers, shocked and angered many in the news department at ABC. But ABC attorneys apparently felt they had no case. ... Indeed, even an FCC lawyer expressed some surprise over ABC's easy capitulation. "We've made it very difficult" for the subject of a news story to intervene, the attorney said, "because of the First Amendment guarantee of a free press." ... Clearly, ABC's original decision was made by lawyers, not journalists, and it was made in an atmosphere of increased corporate activism in the media.*

Kaiser submitted a ten-minute videotape rebuttal, demanding that it be broadcast on *20/20*. ABC News countered with an offer of a four-minute absolute maximum for an unedited rebuttal. After some skirmishing, Kaiser submitted a four-minute tape. ABC News intended to create a new *20/20* segment, in which they would air the Kaiser rebuttal and then pick it apart. Kaiser demanded that its rebuttal must stand as the last word. The skirmishing resumed.

On February 4, 1981, Kaiser issued a press release titled "ABC Reneges on *20/20* Rebuttal". It reiterated their "Trial by Television" arguments and portrayed ABC as having violated a series of written and verbal agreements. The company also filed a formal complaint with the FCC that centered on the concept of "personal attack", claiming that:

> *... These statements, as well as the program as a whole, charged Kaiser with selling a product that it knew to be dangerous, a product*

# 9. TRIAL BY TELEVISION 101

*that has caused numerous deaths. ABC further alleged that Kaiser deliberately concealed evidence of the danger from the government and the public.*

*It is clear beyond dispute that such charges, which were aired during a discussion of controversial issues of public importance, constituted an attack "upon the honesty, character, integrity, or like personal qualities" of Kaiser. It is also clear that Kaiser is "an identified person or group" within the meaning of (Section) 73.1920 of the Commission's rules. ...*

*Kaiser's rights under the personal attack rule do not depend on the truth or falsity of ABC's statements. Accordingly, this letter will not undertake to discuss the accuracy of ABC's charges except to note that they are, in fact, false.*

Resolution eventually came with ABC's creation of a new and short-lived show called *Viewpoint*, hosted by Ted Koppel, which apparently was created specifically to deal with Kaiser's complaint. The show premiered in prime time, 10:00 to 11:00 on July 24, 1981. According to the New York Times:

*Roon Arledge, President of ABC News, spoke of "an opportunity for those with serious complaints about television news coverage to have their views heard, and for us to respond to their complaints." ... "We expect the program ... to help open up better lines of communication between ABC News and those we cover."*

*The more cynical of television observers may suspect that this altruistic project became a bit more urgent and practical after Kaiser Aluminum filed a $40 million libel suit against ABC News in connection with a 20/20 segment broadcast in April 1980. ...*

After exploring issues and problems related to TV reporting, the program's host Ted Koppel introduced the Kaiser complaint as a case in point. The stage was set by showing some scenes from the *20/20* "Hot Wire" segment, followed by Kaiser's four-minute unedited response. This consisted of statements by Kaiser President A.S. Hutchcraft alternating with statements by others. Hutchcraft and the others denied the *20/20* allegations but did not provide any factual support for the denials.

Kaiser's video starts with Hutchcraft posing three questions; "Is aluminum wiring really unsafe? Did we, Kaiser Aluminum, really put a product, a hazardous product, in place? And, did we cover that event up?" His answer to all those questions was "... no, a very definite no." The scene switched to show George Flach, Chief Electrical Inspector for the City of New Orleans, who stated his opinion that aluminum wire "is just as safe as copper when it is properly installed," and that he himself lives in an aluminum-wired house. Flach did not mention that his department specified the use of inhibitor compound for all aluminum wire connections in its jurisdiction. "Properly installed" in New Orleans was not the same as in the rest of the country.

The Kaiser rebuttal video did not correlate well with the company's documented actions and test results. Testing was portrayed, but no mention was made of the results. The failures that occurred were not mentioned, nor was Kaiser's analysis of the fundamental properties of aluminum that caused them. Kaiser's video concluded with Hutchcraft saying, "It is unthinkable that our company would market a product that would put human life in jeopardy."

Then Rivera and Hutchcraft faced off in a debate, moderated by Koppel.

> *... Mr. Rivera argued that Kaiser's demand that its spokesman appear live and unedited on the* 20/20 *report would have required the television correspondent's "abdicating responsibility as a responsible journalist." The businessman, attacked what he called "trial by TV," in which the accuser is also judge, jury and prosecutor. Mr. Koppel managed to remain admirably objective and persistent, to*

## 9. TRIAL BY TELEVISION

*the point where Mr. Rivera may understandably have wondered which side his ABC colleague was on. In the end, the questions were not settled definitively one way or the other, but the viewer was certainly left with a clearer understanding of the issues.*

Kaiser had acted as would be expected of a company that was defending against product liability lawsuits and government action. Protection of the corporation demanded that it could not yield an inch on the basic issues of marketing a hazardous product and failing to warn the public.

The *20/20* piece on aluminum wire had put a National spotlight on the aluminum wiring fire hazard and helped alert the public. For some, however, it was too late.

# 10.
# FATAL FIRE—HOUSTON, TEXAS, 1978

A SLEEPING OCCUPANT IN THE RIVINGTON APARTMENTS COMPLEX IN Houston, Texas was awakened by sounds coming from the wall behind her bed. The noise and a shower of sparks came from an aluminum-wired receptacle. Quick response prevented the situation from developing into a fire. It was a near miss. An electrician replaced the damaged receptacle. No further work was done, since there was no concern about any of the other aluminum wired receptacles in the complex.

The electrician and others involved apparently were not aware of the broader problem. They considered this near miss to be just an isolated random failure. This was due in part to the fact that the industry and UL had not alerted the public to the hazard and had not published any substantive advice regarding corrective action. In fact, they essentially denied that there was a hazard. "Aluminum wiring is safe if properly installed" was the way the denial was, and still is, generally phrased. Further, Kaiser had taken legal action against the CPSC, trying to prevent the safety agency from issuing any advice on the aluminum wiring hazard and corrective actions.

The incident described above occurred at a receptacle in the wall separating two apartments. Dr. Buning, a Houston physician, slept in the adjacent apartment. On the wall, behind his mattress, was an aluminum-wired receptacle, located only a few feet away from the one that—on the other side of the wall—had sizzled and sent out a shower of sparks.

When Dr. Buning failed to show up at his office as scheduled and did not call in, his colleagues became concerned. After trying to reach him by phone, they took action that eventually led to entry into the apartment. It was filled with smoke. His mattress was smoldering, with him in the bed. He had died from carbon monoxide poisoning and smoke inhalation while sleeping. There was no flame, and there was very little structural damage. Investigators determined that an aluminum wire connection failure at the receptacle had ignited a smoldering fire in the mattress. The level of certainty was essentially 100%.

The Buning family—his estranged wife and their children—sued Reynolds, the aluminum wire manufacturer, and GE, the receptacle manufacturer. Representing the Buning family was the firm of Haynes and Fullenweider. The senior partner of the firm was Richard "Racehorse" Haynes, a legendary and colorful criminal defense lawyer who, according to the American Bar Association's Journal, "has been the defense attorney in some of the most prominent Texas murder cases ever tried. He's been memorialized in three books, two movies, a Broadway play, and even in popular music."

There was no dispute as to the point of origin and cause of the smoldering fire. Defendants Reynolds Metals Company and GE argued that their products were safe, met all applicable (UL) standards, and that the receptacle in question must have been improperly installed. They claimed that one or more of the terminal screws had not been sufficiently tightened.

The defense relied on the theories of a British expert, J.B.P. Williamson. He had been the head of Research at the Burndy Corporation in Connecticut, a major manufacturer of electrical connectors. The company hired Williamson in about 1960 to establish a laboratory for fundamental research on electrical connections. He staffed the lab with pioneers in the relatively new field of electric contact technology.

There was much to be learned about the failure processes at the microscopic level of contact and connection interfaces, where one piece of conducting metal presses against another to create a path for electrical current.

## 10. FATAL FIRE—HOUSTON, TEXAS, 1978

Electric contacts and connections were, and still are, considered to be "the weak link in the electrical system."

Most people at the time knew about "bad connections" on the telephone, and many had experience with radios and TVs that would come back to life after banging once or twice on the side of the case. Automobiles would not start—or stalled on the road—when battery or ignition connections failed. Early computers and complex electronic systems were plagued by contact and connection failures that caused "Monday morning blues", when equipment that had been working fine the previous week failed to restart properly after a weekend shutdown.

The timely development of scanning electron microscopes and other sophisticated equipment helped the Burndy researchers shed light on deterioration processes at the contact interfaces. Their effort yielded some ground-breaking insights.

The market for manufactured goods was evolving into one in which lowest price prevails. Corporations were increasingly reluctant to fund expensive fundamental research. Williamson arrived at his lab one morning to find the entrance padlocked. Burndy had abruptly shut it down. He returned to England and started his own consulting business.

The shutdown of Williamson's research lab at Burndy was somewhat analogous to Kaiser's abandoning its own connection testing more than a decade earlier. It had long been common for companies to use their own laboratories, develop their own products, conduct design reviews, and convene company product safety review committees to help assure that their products worked properly and were free of unexpected hazards. The money spent on these activities was readily justified back in a time when brand loyalty and product quality were major selling points. UL testing then provided an added layer of assurance and the label of approval necessary for acceptance in the marketplace.

With the modern trend toward a "survival of the cheapest" marketplace, product safety assurance is often assumed to be adequately covered by UL (or alternate) standards and test procedures. UL in turn, depends largely on the manufacturers for technical expertise in the development of the standards and test methods, and for prompt solution of problems that occasionally arise. For most products, this arrangement appears to be adequate. For aluminum wiring, it was not.

In the Buning litigation, Williamson's claim that the receptacle terminals must not have been tightened properly was contradicted by test results. Under actual use conditions, properly tightened connections were shown to loosen for at least three reasons. First, the aluminum wire itself relaxed under the pressure of the terminal screw. Metallurgists call this "creep" or "stress relaxation" depending on how it was measured. This could account for as much as 50% loss of contact force in a relatively short time. Next, engineers at Kaiser demonstrated that steel terminal screws on receptacles could actually turn and loosen due to a phenomenon known as "thermal ratcheting" caused by ambient temperature changes. Lastly, and perhaps most common, was that the screws loosened when the receptacles were originally installed due to forces imposed by the wire when the receptacles were pushed into the box in the wall. This was clearly shown in Wedekind's test at UL's Santa Clara laboratory.

Williamson argued that none of this mattered if the screws had been really tight in the first place. The industry had—after the fact—come up with a recommended minimum level of tightening; 12 in-lb of torque (twisting force). If the installer did not have a torque-measuring screwdriver, which was usually the case, it was recommended to tighten until snug and then 1/2 turn more. The problem with both of these recommendations is that the screw terminals on many of the available receptacles could not withstand being tightened that much without suffering damage. The screw threads on some receptacles failed ("stripped") at tightening torque substantially below the value that Williamson said was required to prevent loosening and to establish permanent metallic contact through the insulating aluminum oxide.

## 10. FATAL FIRE—HOUSTON, TEXAS, 1978

According to Williamson, the brittle insulating oxide on the relatively soft aluminum metal was like "a thin layer of ice on mud". Under the pressure of the tightened screw the oxide cracked, so that metallic conducting areas were formed by aluminum oozing through the cracks where the aluminum wire and the brass plate pressed against each other. That, he explained, is why it was relatively easy to make a functional initial contact to aluminum in spite of the insulating oxide. What he could not explain, and what had confounded the industry for so many years, was why many aluminum connections, properly made and tightened, deteriorated over time and overheated to hazardous levels.

Williamson coined the phrase "self-healing" to describe the erratic behavior of deteriorating aluminum connections. Overheating aluminum wire connections occasionally dropped to normal temperature for some time. This is what Williamson considered to be "self-healing". He claimed that if an aluminum-wired screw terminal was adequately tightened, self-healing events would continuously prevent the overheating from ever becoming hazardous. This is analogous to remissions for a cancer patient, which are sometimes permanent—but not always. Williamson had no answer as to why the overheating often resumed some time after a self-healing event.

At the Buning vs. Reynolds trial, Williamson's theory did not convince the jury, which awarded Buning's wife and children $1.5 million in damages. Underlying this tragedy was the industry's failure to warn the public and the trade of the potential hazard and to prescribe effective corrective action. The only serious industry effort to do so, through an "ad hoc" (temporary) committee convened by UL in mid-1972, had been effectively thwarted by UL President H.B. Whitaker.

# 11.
# THE UL AD HOC COMMITTEE

Following an industry committee meeting at NEMA headquarters in August 1971, W.H. Abbott, a researcher at Battelle Columbus Laboratories, wrote to UL's VP W.A. Farquhar:

> *In response to Baron Whitaker's letter ... I am forwarding my comments on aluminum wiring. Specifically, the question to be treated is the status of present supplies of wire, wiring devices, and connectors.*
>
> *Based on studies at Battelle-Columbus we believe that under no circumstances should aluminum wire which is presently on suppliers' shelves be connected to conventional binding head screw type devices. ...*
>
> *... we believe that all aluminum in circuit size wires presently in dealers' stocks should be withdrawn from the market. ... connection to devices presently in supplies will inevitably present hazardous conditions.*
>
> *... One final point which was not discussed ... must be considered by the committee. This concerns the question of what will be done about the potentially hazardous connections which are already in existence. ... connection to certain types of wire/device combinations ... may exhibit real time intervals to failure in the period of 5 to 10 years' service.*
>
> *We recognize that this is not a popular subject for discussion.*

> *However, this committee should consider its public responsibility and consider what steps can be taken to avert serious field failures ...*

Abbott headed a team at Battelle that was involved aluminum connection research sponsored by some of the manufacturers. He was well versed in the relatively new specialty of electrical contact technology. Abbott knew how connections worked and how they could fail. From time to time he spoke out in ways that contradicted the positions of his sponsors.

Others working within the industry were reported to have called for UL to withdraw its approval of aluminum wiring. Abbott did it formally, in writing. Since he based his letter on a substantial body of laboratory test results, it had considerable credibility. Abbott's direct message, using words such as "hazardous" and "public responsibility", was unique among the documents circulating within the industry. It added a big wave to the rising tide of aluminum wire bans and adverse publicity that UL was confronting.

UL at that time was focused on improving the performance of future aluminum wire installations. Working toward that objective were several ad hoc committees (temporary and single-purpose) that UL formed to address possible changes to the relevant UL standards. The members of these committees represented the product manufacturers and UL. There was no outside representation.

Nothing was being done by UL or the industry to alleviate the hazard in aluminum-wired homes already in place or in those that would be installed before any product changes were effectively implemented. Abbott's letter added to growing pressure on that issue from both inside and outside the industry. Whitaker and Farquhar eventually convened another ad hoc committee, formed to be a "forum of participants committed to improving safety in electrical systems using aluminum conductors."

Invitations were sent to people on a list created by Farquhar and NEMA Technical Director H.P. Michener. They had previously agreed that the purpose of the initial meeting would be to "assist in putting together an agenda

## 11. THE UL AD HOC COMMITTEE

or list of objectives for a later meeting". Aluminum conductor manufacturers, the Aluminum Association, wiring device manufacturers, NEMA, and UL were represented. The committee agenda and objectives were to be set before anyone outside of the industry would be invited to participate.

The first meeting took place on August 1, 1972 at NEMA headquarters. According to Ed Buja, an attendee from Anaconda Wire and Cable Co., UL President Baron Whitaker ran the meeting and started with a summary of the background of the problem. Buja reported to his company that Whitaker's key opening points were:

- UL had become *"an unwilling referee between wire and connector people"*.

- A member of Congress was making inquiries as to what UL was doing *"in the problem area of aluminum"*, and legislation to ban aluminum wiring was pending in several states.

- *"Trade confusion existed as to how aluminum should be connected, with inconsistent data coming from various manufacturers."*

- *"There is need to make useful and consistent presentations to the trade in this area, and soon, so that the problem is rectified with minimum loss"*.

Buja reported that Whitaker then raised questions as to the role of publicity, whether satisfying the need for better information to the trade was an appropriate function for UL, whether others outside of industry should be invited to participate and whether amendments to the National Electrical Code (NEC) were in order. The attendees agreed that UL was the appropriate organization to "guide" this "coordinating activity" and that other "concerned interests" should be included. A list of additional organizations invited to be represented at subsequent meetings was agreed on. Invitations were sent out by Farquhar the next day.

The people present at the first meeting were also asked to provide a written summary of their publications and activities "in connection with the proper use of aluminum conductors in electrical wiring systems", to be shared with the other attendees before the next meeting.

The second meeting convened on August 17, again at NEMA headquarters in New York. The expanded attendance included representatives of other electrical industry and trade associations and one representative from the National Bureau of Standards. Buja again attended on behalf of his company, Anaconda, and wrote in his meeting report:

> ... *the materials which had been sent to the first attendees were quickly read through. In general, these materials were old specs, meeting notes, or summaries of how the problem was supposedly handled. Copies of these materials are not attached as they number about one hundred pages and are not very useful (glad to supply on request). One would be apt to conclude from these reviews that the various organizations, AA (the Aluminum Association), NEMA, and U.L. had been on top of the problem and that the problem had been properly handled. ...*
>
> *Weitzman of Leviton said that the original introduction of aluminum had caught the device people unprepared.*
>
> *W. Abbott of Battelle ... Steel screws and hard drawn aluminum wire are the worst combination (U.L. accepted practice 1966—1971).*
>
> *Domsitz of the National Bureau of Standards ... The N.B.S. Fire Research Division phone keeps ringing with the reports of aluminum fires. ... Various congressmen have contacted N.B.S. as to the nature of the aluminum problem.*
>
> *E. A. Brand, EEI. (Edison Electric Institute, representing the electric utility companies) Why aren't inhibitors being used? The utilities used them to solve the overhead connector problem.*

## 11. THE UL AD HOC COMMITTEE

*The meeting then entered into a discussion led by Whitaker as to where we go from here. Great concern was expressed about inventories on dealers' shelves ... Most of the technical representatives were so disgusted with the whole proceedings they refused to enter into the discussion. U.L. (Farquhar) said that they were going to make an "emergency suggestion" to prevent the use ... unless pigtails were used. Leviton and Battelle protested that pigtails were the worst thing they could do.*

Whitaker then abruptly ended the meeting. He asked each representative to send him short- and long-range recommendations and said he would appoint a committee to review them prior to the next full session of the "Forum".

During the meeting, the question of aluminum wire's lack of compatibility with the real world of wiring device, contractor, handyman and homeowner installation variables was discussed in some detail. Buja wrote:

*This area of workmanship came up a number of times, but didn't seem to be an area that AA, NEMA and U.L. felt they had any influence or responsibility. Another aspect of this problem is the homeowner who replaces devices in his home. In many cases there is no warning not to use the push-in type on aluminum and this leads to trouble. Schoerner of Southwire bluntly raised the question as to whether conductor people have to make materials that will function properly despite the conditions above. How good is good enough?*

The answer to Schoerner's question is reasonably obvious. Since the industry was marketing aluminum as equal to copper for house wiring, it should perform as well and as safely. In the absence of crystal-clear warnings and special installation requirements, that is what the trade and the public expected and paid for. The poor lab test results and the burnouts and fires in homes amply demonstrated that the aluminum wiring being installed in homes did not meet that expectation. Where special warnings or instructions would be appropriate—regarding the push-in receptacle terminations that Buja mentioned, for example—the aluminum wire manufacturers had consistently and successfully opposed them because they could have a negative effect on sales. UL had the authority—and presumably the responsibility—to require whatever warnings and instructions would be necessary to assure that aluminum wire would be safely used. But the UL executives were reluctant to act on that.

Buja indicated that UL President Whitaker apparently did not believe that UL had any "influence or responsibility" regarding whether the product would perform properly (safely) across the predictable range of installation variables in the real world of home construction and maintenance. That is an understandable position for the Aluminum Association and NEMA, which are industry advocates, but UL has a unique role. The National Electrical Code (NEC) requires that the wiring and equipment be "suitable for the purpose", and it was universally understood at the time that UL had the responsibility to assure the suitability of aluminum wiring in the field of actual use and the authority to halt its use if it proved to be unsuitable.

Concluding his meeting report, Buja wrote:

*It has become evident during the last two meetings as to why U.L. has formed the FORUM* (the ad hoc committee). *Under the concept of creating a group to exchange ideas and information, they are getting the major associations and companies in cahoots with them in essentially a mess that U.L. created. Attempts at leading the discussion in a responsible but unpleasant direction are cut off and we deal with trivia. U.L. is continuing to shirk their responsibility and I don't feel we should support the direction they want to take …*

Whitaker steered the committee toward the objective of developing an

## 11. THE UL AD HOC COMMITTEE

information package that would proclaim—especially to critics outside the industry—that the problems had been addressed. Three items constituted the centerpieces of his information program: a) an information pamphlet for general distribution; b) a "stuffer" for the utility companies to distribute with their monthly bills; c) an informative press release.

From the end of September 1972, when the first draft of the pamphlet was sent to the committee, to mid-March 1973, an enormous amount of industry effort was spent reviewing and revising it, proposing changes and writing memos. The full committee itself met five times during that period, with an average of about twenty attendees at each meeting. The ad hoc committee spawned its own subcommittees, in addition to numerous temporary committees and subcommittees within the various industry organizations. Top management and legal departments of the companies and the various organizations were involved.

Alarmed by the information that was surfacing at the time and unable to change the path down which UL was taking the industry, some companies took action on their own. Hatfield Wire and Cable halted production of all aluminum building wire products at the end of October, 1972. T.J. Stewart, Hatfield's Vice President-Marketing, sent the following explanation to the company's sales force:

> ... *The products withdrawn are used primarily in residential and commercial construction. More and more, we have been coming to the conclusion that aluminum conductor presents serious problems in terminations and splices. These problems have led to code restrictions in some areas due to the fire hazard that exists. American Plan Insurance announced in the Wall Street Journal that they were not going to insure mobile homes wired with aluminum conductors due to fire losses. Panel box manufacturers report problems and wiring device manufacturers withdrew their products for use with aluminum conductors.*

> *If all these circumstances are not enough, reports by an independent laboratory retained by the aluminum industry now reveal that its failure incidence at terminations or splices increases after being in use for several years. This "time bomb" effect, coupled with our responsibility as a good corporate citizen, makes it a moral obligation to take this position. If at some future date a responsible independent testing authority can re-establish our confidence in this product, we will reconsider our position at that time. ...*

By March 1973, after more than a half year of effort, the ad hoc committee had produced a six-page brochure (plus front and rear cover), a final version of the stuffer for the utilities and a "safety message" press release. The time and effort expended toward review and refinement of such seemingly straightforward items reflected the extreme sensitivity of the issue. By that time, more than two million homes had been wired with aluminum. The proposed public statements touched on substantive and potentially costly issues. It was a public safety issue. On the line were UL's credibility, corporate product liability and the future sales of aluminum wire. The market for aluminum wire could collapse. Product lines, jobs and careers were at risk. Companies might have to pay for corrective measures.

UL President Whitaker presided at the ad hoc committee meetings and personally wrote somewhat whitewashed summaries—in lieu of detailed minutes—that were sent to the participants. The attendees' own meeting reports, written for distribution inside their company or organization, demonstrate that there were serious objections to the main points covered in the proposed pamphlet.

The final (published) version of the pamphlet implies that improper installation was at the root of the aluminum wiring problem, that the problem had been solved with the new UL listed "alloy" aluminum wires and the "CO/ALR" wiring devices, and that existing stocks of the older aluminum wire could be safely connected directly to the CO/ALR devices

## 11. THE UL AD HOC COMMITTEE 119

or to standard devices using the pigtailing method. None of this was strictly true. Those involved who actually had lab test data and field experience argued strenuously against making these claims. Whitaker dealt with the objections by asking that details be put in writing and mailed to him or Farquhar. Some of the responses that were received were distributed to the members of the committee. The most critical were not.

Battelle's W. H. Abbott challenged, among other things, the "improper installation" concept, writing as follows:

> *This document gives the reader the impression that most of the "blame" for field problems should be placed on the installer. ... all available information indicates that the problem is a systems problem of device, wire, and installation. In fact, other than the case of totally loose connections, evidence indicates that the basic installation problem/factor may be one over which the installer has little practical control.*

The bulk of the objections centered on the suggestion that the pigtailing method should be used to connect aluminum wire to anything other than the newly-tested CO/ALR type devices. UL had not done any substantive testing to confirm that the twist-on connectors that it listed as suitable for that application would actually provide safe long-term service. Members of the committee recommended against pigtailing based on their own experience. Buja, in his usual meeting report to his company, wrote:

> *It was a tedious and stormy session, first rewriting a 20 page statement prepared by U.L., and then a conflict developed on a technical point—should pigtailing 1 copper and 2 aluminum conductors be recommended by the group? Whitaker was obviously upset by this, and the group refused to be convinced by U.L.'s arguments that*

*pigtails were OK as there is absolutely no U.L. Lab tests, and there has been some bad experience on the part of some participants.*

Whitaker, in his report (minutes) of that meeting addressed to the committee members, wrote:

*The concept of "pigtailing" copper to aluminum conductors received considerable attention. Limited testing under the same program as used to evaluate the new "CO/ALR" devices had produced unsatisfactory results yet there is apparently no evidence of field failure where the pigtailing practice has been used for the past two years and is in fact required in certain localities. The wire connectors used in pigtailing had performed satisfactorily in laboratory tests where subjected to ... smaller overload currents than had been used in ... tests of wiring device terminations. It was generally agreed that the evaluation of connectors with respect to long-time reliability should be more thoroughly explored. On the other hand, there was a counter feeling that considering the good field record to date, the use of such connectors for pigtailing should not be prohibited. The current text reflects the general consensus of views on this subject.*

The "counter feeling" that prevailed reflected the fact that the majority of the committee members—Whitaker's "general consensus"—represented the aluminum companies' interests one way or another. Whitaker had written early on that there was "agreement" that the committee would develop instructions for continued use of the older EC aluminum wire in stock, and pigtailing was the only available solution. The aluminum wire manufacturers faced substantial financial losses if the existing stocks could not be sold.

J.F. Tibolla was chairman of the General Engineering Committee of the NEMA Wiring Device Section. He represented that section on the ad hoc committee, and reported to NEMA that:

## 11. THE UL AD HOC COMMITTEE 121

*I am in general agreement with the main body of the statement with the exception of ... paragraph two of Page 4 regarding the use of pigtails. In the limited testing that we have performed at Circle F Industries, we cannot sanction the use of pigtails with pressure wire connectors as a satisfactory interim solution for using up existing inventory. ... I asked that the UL or that the wiring connector people submit data to the contrary and this has not been done.*

Adding to the objections, R.O. Wiley, chief engineer at Bryant, a wiring device manufacturer, went on record as a NEMA committee chairman, writing that:

*From the industry's past experience with aluminum wire, I believe that NEMA should not be a party at this time to a statement which encourages aluminum conductors as branch circuit wire. Even though apparent improvements have been made ... there is not yet any basis for determining that any terminations designed to date will, in fact, provide safe, long term results in field use.*

*It appears that there is considerable pressure to issue a statement without regard to all the facts. In fact, in this whole problem of aluminum wire, it seems that many facts have been pointedly ignored. For instance, it has been the experience of this company and I am sure many others, that the twist-on or "wire nut" type of connector is not entirely satisfactory for use with aluminum, and they are being recommended for use regardless ...*

*I feel it is better to issue no statement at all, than to issue a statement which may tend to increase the use of aluminum wire and possibly result in higher instances of field failure than have been experienced to date. Therefore, if this is the type of statement which UL intends to publish, I feel that NEMA should not have its name included as one of the issuers of this statement.*

A sort of rebellion had been brewing on that last point. Whitaker had written:

*It is hoped that each member of the Ad Hoc Committee would be agreeable to being listed as a member of the Ad Hoc Committee thereby indicating broad endorsement of the presentation.*

It was not to be. B. Falk, the NEMA President, did not want NEMA to be associated by name with the ad hoc committee pamphlet. According to H. Leviton, none of the individuals or companies wanted to be named. The participants wanted UL to own up to its authorship. Leviton, President of one of largest wiring device manufacturing companies, wrote to his chief engineer, M. Weitzman, that he did not like "being party to a resolution which is nothing more than a face-saving White Paper for the aluminum people and the Underwriters' Laboratories." The Aluminum Association did not want its name on the pamphlet because it singled out aluminum wire problems.

The pivotal stumbling point was the issue of pigtailing. UL was heading down the same road that previously led to their present problems. It was recommending an aluminum wire termination method without having any test data to confirm its safety. The people, companies and organizations that Whitaker wanted to endorse the pamphlet with its pigtailing recommendation were wary, and rightfully so. Many of them had experienced failures of aluminum wired twist-on connectors used in their receptacle testing setups. Whitaker countered with his statement that the field experience was good.

Whitaker's claim was based on a lack of data, rather than a thorough investigation. There was no failure incident reporting or investigation scheme in place by which UL or any other individual or entity could possibly have gathered substantive data to support his claim. Whitaker and Farquhar certainly knew that the claim had a shaky foundation. For several years, UL had been receiving reports of burnouts of twist-on connectors

## 11. THE UL AD HOC COMMITTEE

used with aluminum wire. Farquhar himself had been involved in some of that correspondence. Only a year earlier, UL sent out a Bulletin that noted reports of overheating of "wire nut" connections at aluminum-wired hot water heaters.

Several days after that meeting, Farquhar sent a letter to M.G. Domsitz, the NBS representative on the ad hoc committee, saying:

*In view of some of the discussion which took place at the meeting on November 10th, Underwriters' Laboratories is anxious to try to obtain factual information from the field regarding experience with binding screw terminations and pigtail terminations. We have assembled two questionnaires, which we are proposing to send to chief electrical inspectors throughout the country with the hope that they will take the time to respond in a meaningful manner. There are approximately twelve hundred of these on the mailing list.*

*… I am not taking the time to send this to each of the members of the committee, but rather would like to know whether you, based on your experience in this area, would suggest any changes in the questionnaire …*

Domsitz had been Chief of the Electronic Technology Division at NBS. The most likely reason that Farquhar quickly alerted Domsitz to the proposed survey is that NBS was known to be in contact with the Congressional committee that was investigating the aluminum wiring problem. About a month before Farquar sent his letter to Domsitz, L.M. Kushner of NBS had sent Representative J.E. Moss a summary of the status of aluminum wiring as known to NBS at the time. Kushner indicated that the industry was moving in the right direction to solve the problems. That letter kept Congress at bay for the moment, and Farquhar's special attention to Domsitz was most likely designed to help fend off a resurgence of Congressional pressure.

After Domsitz had a chance to review the questionnaires, Farquhar sent them to the other committee members who, in turn, sent them out for review by their various companies and organizations. On the day after Christmas, the final version, a single questionnaire, was sent to the electrical inspectors' organization, IAEI, to be mailed to the 1100 or so Chief Electrical Inspectors on its list.

According to Whitaker's summary of the next meeting, the ad hoc committee agreed to proceed with the pamphlet as proposed, including the suggestion of pigtailing, unless the inspectors' questionnaire returns indicated the need for change.

There was no reason to believe that abnormal overheating or burnouts of the splices would come to an inspector's attention. The electrical inspectors' role is to check that new residential construction installations and modifications to existing installations conform to the codes and standards that are applicable in their jurisdiction. Problems that show up after the original installation is completed and inspected are generally handled by electricians, handymen or homeowners without an inspector being involved.

There are no requirements to report overheating failures and burnouts of receptacles and connectors to an electrical inspector—even if they pose a fire hazard. If the fire department was involved, the local electrical inspector might or might not be called. A charred or burned up receptacle or splicing connector is most likely to be just replaced and thrown in the trash. The responses to the questionnaire reflected these facts. Most respondents had no data to offer.

Whitaker's survey did not have much chance of zeroing in on a well-defined population of homes with pigtailed aluminum wiring, and the electrical inspectors were not likely to be aware of failures and burnouts that did occur after installations were completed. Further, none of the pigtailed aluminum wire installations were more than about two years old.

In spite of these limitations, the results were significant and confirmed the concerns that had been voiced at the meetings. The compiled pigtailing

# 11. THE UL AD HOC COMMITTEE

data of the questionnaires shows 107 known pigtail splice failures. UL received a total of 475 responses out of the 1100 that were sent out. Only a portion of the inspection departments that responded knew of any aluminum wired homes in their jurisdiction. Only 17 knew if any of those were pigtailed. None of the respondents knew if the method had been used in mobile or manufactured housing in their region. Considering the limitations of the data gathering method and the short time that the pigtailing practice had been in use, the report of 107 pigtail splice failures sounded a clear alarm.

The results were discussed at the February 23, 1973 ad hoc committee meeting. According to S. Kolmorgen, who represented Anaconda in place of Buja:

> *D'Agostino gave a verbal report on the UL survey on aluminum termination problems and pigtailing field experience. It is not too conclusive because of poor response (less than 50%) and the fact that it is based on memory of electrical inspectors and not written records. Mr. Whitaker had Mr. Farquhar try to summarize these figures but other than add some generalities of his own it was not too impressive.*
>
> *Mr. Whitaker concluded the discussion by stating that the survey did not disclose a panicky or runaway situation but they needed to keep looking into it, he intends to take the written report figures which he would not hand out because he felt they were more confusing than helpful. He will revise them to make them more creditable before handing out. He doubted that publication of the survey would occur before the pamphlet ... is printed and issued. ... He stated that they would print up 18,000 copies ... the pamphlet would be printed in 30 days and would be mailed out immediately thereafter.*

Whitaker did not abide by the agreement made at the previous meeting, which was that the pamphlet with its pigtailing recommendation would not be published if the survey results indicated field failures where the method had been used. He essentially kept the members of the committee in the dark as to the results of the survey until it was too late for any of them to protest or stop the pamphlet's publication. Whitaker apparently did not consider that the reported 107 pigtail splice burnouts signaled a problem.

Whitaker decided that there was no need for additional meetings of the ad hoc committee. There was to be no additional group discussion and consensus. Whether or not the members of the committee agreed that pigtailing aluminum wire was safe no longer mattered. UL moved along with publication and distribution of the pamphlet, including its promotion of pigtailing.

The UL pamphlet, titled *The Use of Aluminum Conductors With Wiring Devices in Electrical Wiring Systems*, was issued and sent to people in the electrical industry and trades, the insurance industry, and to anyone who requested it from UL. It was sent out to only a few homeowners—those who contacted UL with inquiries or complaints related to aluminum wiring.

The pamphlet's recommendation of pigtailing is implied, not explicit. There is only the statement "Pigtailing … as illustrated in Figure 1 is recognized by the NEC". That suggests that the practice is safe, contradicting the data available at the time. Once again, UL had not done the testing needed to assure that the twist-on connectors that they listed would perform safely in that service; that they were actually "suitable for the purpose" as required by the NEC.

At the same time that the UL pamphlet promoting pigtailing made its debut, S. Myers, Mayor of Buena Park, CA, a retired Fire Prevention Captain from Santa Fe Springs, testified at the CPSC hearing in California. He stated that a new policy requiring pigtailing was established in 1970, after some fires caused by aluminum wire termination failure. Subsequently the pigtail splices were found to be failing.

Since that time, burnouts and related fires involving aluminum wire pigtail splices have amply demonstrated that the practice is not necessarily

## 11. THE UL AD HOC COMMITTEE

safer than connecting the aluminum wire directly to screw terminals. Made with the commonly-used UL listed (for the purpose) twist-on connectors, pigtailed aluminum wire connections fall far short of the equivalent safety of copper wired connections.

A 1996 field report, for instance, states that there was:

> ... a small fire in my parent's mobile home ... and concern about the safety of their remaining in the home after repairs were made. Two outlets had already failed in the past and I found that one of the [pigtailed] repairs had heated to the point that one of the wirenuts had melted and charred insulation.

And another, from 1999 states:

> I am a retired electrician after over 40 years service ... My Daughter just purchased a home in Chalmette, LA near me. She needed an added electrical outlet in the pantry and I investigated an existing outlet in the kitchen. When opened I saw that the wire nuts had melted and the insulation was charred and burned back a bit. What more amazed me was the fact that the house was re-done with copper pigtails that were recommended years ago. It seems that this remedy was a failure ...

Whitaker did not want UL to be too closely associated with this pamphlet. Sending copies to various individuals and organizations, he did not use UL stationary for the cover letters. Instead, he used stationary with the letterhead "Ad Hoc Committee on Aluminum Conductors In Electrical Wiring Systems". He signed the letters without identifying himself as UL President. In small print, the letterhead identifies Whitaker as Chairman of the Committee, with a UL address but without mentioning his prestigious title. In the following years, UL sent the pamphlet as a standard response to homeowner inquiries, with cover letters on this same Ad Hoc Committee letterhead.

There is no statement anywhere in the pamphlet that any person or any organization or company has responsibility for the pamphlet's content. The only text on the pamphlet's cover page other than the title is, "A statement prepared under the auspices of an Ad Hoc Committee sponsored by Underwriters' Laboratories, Inc.". The pamphlet's back cover page has only the message "Representatives of the following organizations served on the ad hoc committee on the use of … aluminum conductor …", along with a list of the organizations represented on the committee and a UL address from which to order copies at 10 cents each.

Commenting on one of the early drafts of the pamphlet, Abbott had written:

> *As a general statement this document does not appear to offer anything new, does not set forth any affirmative course of action, and is somewhat misleading by omission. Furthermore, it is not evident that this document will serve the educational function considered necessary by most of the committee members. … I seriously question whether this particular document would ever get to the consumer and if it did whether it would mean what it should to him.*

Abbott was right. The key message—*that only the tested and newly-listed receptacles marked CO/ALR should be used with aluminum wire*, and that *devices marked AL-CU should not be directly connected to aluminum wire*—did not get through to people who needed it. To illustrate the point, in 1975, mobile home manufacturer Fleetwood responded to a woman's letter regarding the fire hazard posed by aluminum-wired receptacles that were burning out in her parents' mobile home. The company's reply letter advised her to make sure that the receptacles were marked AL-CU and, if they were not, to replace them with new ones marked AL-CU. The

## 11. THE UL AD HOC COMMITTEE

company, one of the largest in the country (presently on the Fortune 500 list) apparently had not yet gotten the message, some two years after the pamphlet's publication by UL.

Abbott's comments had done little to sway Whitaker from his course. The UL President's stated objective—to clear up the confusion in the trade as to how to "properly" install aluminum wiring—was, in Whitaker's view, achieved by this pamphlet. He had managed to cast a glow of legitimacy over the pamphlet by falsely implying that broad participation in its creation equaled broad agreement with its contents.

The final version of the pamphlet was not widely distributed to homeowners. And, as Abbott had anticipated, it was not of much use in preventing hazardous connection failures in existing aluminum wired homes. That purpose, by Whitaker's reasoning, was to be served by a safety message "stuffer" that homeowners would receive in the envelope with the utility company's monthly bill, in combination with a media press release.

The aluminum interests on the committee were dedicated to watering down or eliminating any message in the stuffer or press release that singled out aluminum wiring as a special hazard. Others on the committee countered that as best they could, advocating for the inclusion of strong language to alert the homeowners and provide effective hazard reduction advice specific to aluminum wiring.

The electric utilities did not want any part of it. The stuffer was never sent out to their customers.

It remained for the press release to do an effective job of informing homeowners. The text had been finalized and agreed to after multiple review and revision cycles. Aluminum wire manufacturers, represented on the committee by the Aluminum Association and NEMA, in addition to their own company representatives, constantly pushed to weaken or eliminate any warnings specific to aluminum wire. The pressure was not totally effective. The agreed-upon final version of the ad hoc committee's press release, although brief and watered down, focused on the aluminum wiring problem and had some useful information toward reducing the hazard.

Whitaker then met with public relations representatives of the Aluminum Association and created a new version of the press release. The wire manufacturers did not want the public to panic and cause their aluminum building wire sales to collapse. Whitaker agreed and personally worked directly with them to create the new version. Now even NEMA objected! NEMA President B. Falk wrote that Whitaker's proposed changes "distorts the original intent of the safety message," and urged that the release be identical to that which the ad hoc committee agreed to, "without changing a comma."

UL President Whitaker held firm. The version that was finally released does not even mention aluminum wire in its first paragraph. That paragraph sets the stage for the rest of the article. It does not catch the eye as a warning message for owners and occupants of aluminum-wired homes.

Whitaker also accepted the Aluminum Association's proposal that it would handle distribution of the press release through Derus Media Services, a company that it did business with. The release was identified as coming from Derus, not from UL or any respected authority or organization in the electrical industry. To a newspaper editor receiving it, the release was not an important consumer safety announcement: it was just another unsolicited article from a PR firm hoping that the paper would use it to fill some column inches of empty space.

Whitaker and the Aluminum Association also severely limited the distribution of the neutered "safety message" press release. They did not release it nationally. Whitaker and the Aluminum Association specifically instructed Derus not to send the release to the country's major newspapers and national wire services. Instead, it was to be sent only to local publications in and near the jurisdictions where the inspectors had reported 50 or more aluminum wire termination failures on the UL questionnaire returns.

At the time, there were about three million aluminum wired houses

## 11. THE UL AD HOC COMMITTEE

and mobile homes across the country. Many ad hoc committee participants thought that an effective safety message should go out to all of them. Whitaker took an opposite approach. He wrote to the ad hoc committee members that:

> ... the Safety Message ... was sent to the Derus Media Service with instructions concerning its release. ... Derus released the Safety Message on August 30 to 766 publications located in 302 cities and towns in 22 states ... The release coverage was quite good reaching into all but eight of the 43 communities identified as having 50 or more failures. ... The actual publication of the Safety Message is optional with the publishers, hence, we do not know at this time the extent to which the Message has actually been used.

The watered-down "Safety Message", for whatever good it might have accomplished, was not distributed at all in the majority of the states, including New York and Georgia, where the Hersh and Johansen fires occurred (see previous Chapters 2 and 4). Nor was it released in the Cincinnati, Ohio/ Northern Kentucky area or in Houston, Texas, where the Beverly Hills Supper Club and Buning fires occurred (Chapters 1 and 10). One can only speculate as to whether a more effective nationally distributed message would have helped avert these and other tragedies.

It is obvious that broad coverage was deliberately avoided. Logically, to alert as many of the people who were at risk as possible, UL should have issued any such release itself to the wire services and national media. Under its own prestigious name, it would have gotten proper attention.

UL and the industry attempted to convey the image that they had the aluminum wiring problems under control. They were not successful. Many legislators were under pressure from constituents to take action on aluminum wiring. More state and local bans were enacted across the country.

At the National level in Congress, the aluminum wire fire safety issue was a major factor leading to the establishment of the Consumer Product Safety Commission (CPSC). Six months before the ad hoc committee had its first meeting, Congressman J. Moss was questioning the continued availability of aluminum wire. His home state of California was buzzing with aluminum wiring connection burnout and fire reports. Representative Moss was one of the originators of the Consumer Product Safety Act, which was signed into law by President Nixon in late October 1972. That act established the CPSC.

The Hersh fire occurred six months after Whitaker wrote his report on the distribution of the press release. That fatal fire—its cause clearly linked to an aluminum-wired receptacle—was the focus of a detailed on-site investigation by the newly-formed CPSC, and it was instrumental in propelling the agency toward a full-scale inquiry into the aluminum wiring hazard.

# 12.
# THE CONSUMER PRODUCT SAFETY COMMISSION

In an affidavit to the CPSC, M.P. Barclay recalled the fire that destroyed her 1968 Parkway mobile home as follows:

> ... It was about 6:00 P.M. on January 7, 1976 when the fire started, I was in my bedroom which was the master bedroom in front, I had the stereo on and had been napping. I heard a sizzling and popping sound and looked over where the stereo cord was plugged in and saw flames shooting out from the outlet. The fire shot out and upwards from the outlet igniting the window curtains. I rushed over to the outlet and pulled the stereo cord out of the outlet and then went across the room and pulled the cord to the portable heater out of the outlet it was plugged into. In a few seconds the entire room was in flames. I ran into another room where the telephone was and called for help, the fire department came in a few minutes. My two sons and our dog were in the rear bedroom and got out of the burning structure safely. ... most all of our of our personal belongings were destroyed by the fire.

Her mobile home was aluminum wired. It was totally destroyed. She had lived in it for four months prior to the fire and during that time had not noticed any electrical problems.

The previous owner also provided an affidavit to the CPSC, stating:

> ... I am familiar with general repair work around a house and farm including electrical repair work such as replacing switches and outlets. ... I lived in the mobile home for 2 years ... I did notice a problem with the wiring ... An outlet in the bathroom smoked and was not functioning properly ... I disconnected the wires from the outlet and taped the wires ... I also noticed that an outlet in the rear bedroom showed signs of overheating because it was discolored, but I didn't replace it because we didn't use it. ...

The previous owner did not realize that the overheating receptacles signaled an endemic hazard in the home's aluminum wiring. He was face-to-face with evidence of the "time bomb" that Hatfield's VP of Marketing wrote about to his company's salesmen. The UL ad hoc committee's "Safety Message", for whatever good it could have done, had been squelched, and the UL-Aluminum Association press release had not been distributed to the media anywhere in Idaho, where the mobile home was located.

CPSC investigators went to aluminum wired homes and obtained scores of similar first-hand affidavits. In contrast, UL and the industry worked on aluminum wiring problems at arms-length distance from the people who were living at risk with the hazard. Most UL and industry committee members who were involved appear never to have seen or handled overheated and burned out aluminum-wired devices and splices. There is no indication that photographs or actual samples of charred receptacles and connectors from homes were ever circulated at any meeting of the various UL and industry committees. Nor did the participants reach out to home occupants who were dealing with the problem first hand.

UL and industry meeting attendees discussed and wrote about aluminum wiring in words that skirted the issue of fire hazard. They used words like "problems", "poor connections", "troubles", "improper connection", "failures", "difficulties", "overheating", and "abnormal overheating". These ambiguous words did not reflect the nature of the beast that they were

## 12. THE CONSUMER PRODUCT SAFETY COMMISSION 135

dealing with. The result was that within the industry there was confusion, conflict and chaos rather than the informed consensus and sense of urgency portrayed by UL and industry's public statements.

In February 1972, about the same time that the industry was proclaiming to the public that the new "alloy" aluminum wire and CO/ALR wiring devices had solved the problem, NEMA President B.H. Falk sent a letter addressed essentially to the organization's entire membership, writing:

*Gentlemen:*

*I need your assistance in resolving a very important matter.*

*Over the past four to five years, reports of field problems with aluminum conductors and terminations have been increasing. These reports come from many sources: electrical inspectors, contractors, … and now the Federal government.*

*NEMA has been active. However, the total effort over these four to five years by individual product areas of NEMA, sometimes in cooperation with outside bodies, has failed to fully identify the nature of the problem, be it material, mechanical, device, workmanship or a combination of all three. I believe it would be difficult indeed to explain to any interested party how this could be—in view of our industry's technological expertise and know-how.*

*The time to act is almost past. Some local and state prohibitions against the use of aluminum building wire and terminations are already in effect, and more are contemplated. A very large segment of the electrical industry is affected. …*

*I have asked the NEMA Codes and Standards Committee to adjust its order of priorities, and to take immediate steps to bring the concerned product areas together in a concerted effort to identify the nature of the problem, the current test programs, and to recommend an industry program aimed at resolving the field problems.*

*The purpose of this letter is to alert you to the need for NEMA to take positive action; and most of all to seek your cooperation and your company's involvement in this NEMA program. ...*

Falk's call for help did not inspire any substantive progress toward alleviating the aluminum wire hazard for homeowners. His letter does not mention fires and fatalities caused by the "field problems". Nor is there any suggestion that the self-regulating industry that NEMA represented had a public responsibility to take effective corrective action. No new research projects were initiated as a result of Falk's appeal. The parties within the industry continued to lock horns over who was to blame and who was entitled to see the results already obtained from the lab testing that was under way.

A prime example involves Abbott's manufacturer-funded work at Battelle. Various research tasks had different sponsors. Contracts stipulated that access to the results was limited to and controlled by the sponsors. The research results were not shared with other entities in the industry unless they paid to join the sponsoring groups and were allowed to do so by the existing sponsors. The reports that Abbott produced for the connector group, which noted poor test results with the pigtail splices, were not shared with the aluminum wire group or with UL.

UL, which—in the public eye—was responsible for public safety on this issue, did not have access to the fruits of the Battelle work. Eventually, six months after Falk wrote that letter, three UL engineers visited Abbott's laboratory and got a brief overview of some of the test work. But the detailed test results were not shared. The poker game that was being played with the research work and test data, all on behalf of the proprietary interests of the sponsoring manufacturers, inhibited timely resolution of the industry problem that Falk referred to.

Falk's letter did elicit some responses. One such was a letter written by F.L. Gelzheiser, Engineering Manager at Bryant, a subsidiary of Westinghouse that manufactured receptacles and other wiring devices. Two

## 12. THE CONSUMER PRODUCT SAFETY COMMISSION

weeks after Falk wrote his letter, Gelzheiser went on the record at the meeting of a NEMA Joint Section Committee on terminations for aluminum conductors. He noted that the committee had been meeting for the past five years and, in his words:

> *I feel that in all these meetings we have skirted the true problem; that of the aluminum wire. Just recently I was shaken by the return of one of our 5501 push-button switches which we have been manufacturing since 1923. The complaint was a failure of the terminal when used with aluminum wire. Did the terminal or wire fail?*
>
> *I feel it is time that we must honestly and sincerely face up to this problem before the United States government, congress ... or any other federally subsidized agency begins to investigate this area. ...*
>
> *At the present time I would be hard put to defend my corporation ... if I were an aluminum wire manufacturer. ... Wiring devices ... had been on the market for endless years and successfully terminated with copper wire with a minimum or no serious problems. With the introduction of aluminum wire, no provision whatever was made to match characteristics of the wire with the product with which it would be used. This resulted in the many problems you are all aware of, and trying to solve by today's meeting. ...*
>
> *Gentlemen, I feel very strongly in this matter and would like to go on record in the minutes of the meeting with this letter.*

One week later, Gelzheizer wrote an internal memo stating that:

> *... I planned to personally present this letter, but could not make the meeting. My representative read the letter and asked that the letter go on record in the minutes of the meeting. Later Mr. Andy Farquhar, of Underwriters Laboratories, called and requested that*

*we do not record the letter, since it could be very damaging. I agreed and withdrew the request from NEMA.*

Two years later, Falk testified on behalf of NEMA at the CPSC hearings in Washington, D.C. His prepared statement had originally been drafted by NEMA's wiring device section. By the time it had filtered through the association's review and revision process and was re-written by the General Engineering Committee (GEC), it had a completely different message. One author of the original draft, Circle F Industries' engineering manager J.F. Tibolla, wrote that:

*The final GEC version appears to me to be a completely watered down version of what the Wiring Device Industry's true position on this matter is. ... It is expected that Mr. B. Falk, President of NEMA, will read this bit of trash in front of the commission ...*

The prepared statement that Falk read at the March 28, 1974 CPSC hearing stated that installation workmanship was the key problem, that the failure mode of aluminum terminations was "overheating in excess of present Underwriters' Laboratories limitations" and that the information available at that time did not lead NEMA to conclude that aluminum wiring presented a substantial hazard.

After reading the prepared statement, Falk concluded his testimony by stating that the NEMA field representatives who are in contact with electricians and inspectors had not reported any significant problems. That was blatantly false. Since at least as far back as 1966, NEMA field men had been reporting an epidemic of connection problems, including some fires, involving a broad spectrum of aluminum conductor applications.

Later that same day at the CPSC hearing, homeowner M.Breheny testified about his personal experience living in an aluminum-wired home in Medford, NY, about 20 miles east of UL VP W.A. Farquhar's office at

## 12. THE CONSUMER PRODUCT SAFETY COMMISSION

UL's Melville facility. Breheny showed the panel two severely charred aluminum wired receptacles from his home. That was persuasive. If Falk stayed in the hearing room after his testimony, that was most likely the NEMA President's first exposure to a person actually living with the aluminum wire fire hazard and his first view of what a receptacle "overheating in excess of present Underwriters' Laboratories limitations" really meant in the wall of a home.

An industry observer at the hearings reported that:

> *... Mr. Breheny's testimony was the most damaging to industry, the testing laboratories and the use of aluminum wiring. He was youthful, angry, and articulate and had done his homework well. His questions to the panel for prompt action were realistic from a consumer's point of view ...*

The observer's report notes that F. Barrett, Executive Director for the Commission, reacted to Breheny's testimony as follows:

> *... up until this point in the testimony by other witnesses, he had begun to conclude that aluminum termination failures were more of a nuisance than a hazard. Now, however, in view of the testimony by Mr. Breheny, Mr. Barrett said that he believed a hazard does exist ...*

(About one month after he testified at this hearing, Breheny alerted the CPSC to the fatal Hersh fire, giving the agency the ability to investigate the cause before the scene was disturbed.)

At the CPSC hearings, testimony by the manufacturers and Battelle often referred to their testing activity. The details and data remained tightly guarded. They were not generally shared among the involved parties within the industry, much less provided unedited to the CPSC.

As an example, a month before the March 1974 CPSC hearings, J. Rabinow of the NBS, who was developing information for the CPSC, met with three UL engineers who described their aluminum wire test methods. Rabinow followed up on that meeting by writing to UL's W.A. Farquhar, thanking him for the cooperation and asking for the data that UL had developed. Farquhar responded by letter, writing:

*We will be glad to provide you with the information you have requested and I am instructing Messrs. Coffey, D'Agostino, and Krawiec that they should do so.*

The three people mentioned are the engineers who had met with Rabinow. E.J. Coffey was the manager above the other two. On Coffey's copy of this letter Farquhar wrote a PS, stating:

*I talked with Baron Whitaker about this ... and have his authorization to provide any of the information which is in shape to be given to the Bureau. Would you please have this done, taking care to see that there is nothing in it which should not go out of the Laboratories.*

The information that the CPSC gathered in the 1974 time frame painted a picture that mirrored Falk's February 1972 letter. Serious problems existed with the aluminum wiring already installed, and the industry was still uncertain as to the scope of the problem and the underlying causes after years of somewhat uncoordinated effort. Congress gave the CPSC the mission of resolving the problem in the interest of public safety. It became clear that the agency could not depend on UL and the industry to provide the information needed to fulfill that mission. The industry had no reliable first-hand information on the field failures.

The CPSC followed up on newly-received field incident reports,

## 12. THE CONSUMER PRODUCT SAFETY COMMISSION 141

sending specially-trained staffers to conduct interviews, collect evidence and determine if failure of an aluminum wire connection actually caused the reported incident. One investigation report concerns a mobile home fire in Oxnard, California. The damage was limited, unlike the total destruction of the Barclay mobile home. Whereas the conclusion as to cause of the Barclay fire relied heavily on the one eyewitness account, the cause of the Oxnard fire cause was solidly established by multiple eyewitness accounts reinforced by obvious physical evidence.

CPSC field investigators from the Los Angeles office arrived on the scene only a few hours after the alarm was called in to the Ventura County Fire Department on the morning of May 7, 1976. The mobile home had not been moved since it was first delivered new in 1969. According to the CPSC report, the owner reported that after the first three to four years in the home:

> *... he began to smell burning odors, and to see smoke coming out of the various electrical switch and receptacle boxes. He said that, up to the present incident, this must have happened 5 or 6 times. He said that what he usually did when this occurred—based on the advice of an electrician friend—was to open up the boxes and tighten the screws on the switch or receptacle which was causing the problem. He said that once he opened up a box in which there were 2 switches on one side and a duplex receptacle on the other—because it was emitting a burning odor—to discover that the insulation of the wire leading into the box was completely gone, and that the wires were naked for more than 12 inches into the wall. ... he replaced the wire by splicing in another piece. ...*

After breakfast on the morning of the fire, the husband had just gone out to the back yard. His wife was in the bedroom and noticed the light flickering. Then there was a sound "like a motor starting up as the lights blinked some more." The CPSC report states that she went back into the kitchen, and:

> ... she saw smoke billowing from the kitchen walls. She ran to the phone and called the fire department. Then she ran and caught her husband who was just leaving. He ran to the main switch and turned off all the electricity. Then he grabbed a crowbar and proceeded to rip off the aluminum siding. He said that when the first pieces came off, he was greeted by a sheet of flame which jumped out at him, but he was unhurt. His wife, meanwhile, turned the water on and sprayed the walls, as he uncovered them, through a garden hose. By the time the firemen arrived, they said, they had put the fire out. ...
>
> The husband is, with the help of his electrician friend, now replacing all of the aluminum branch circuits in his mobile home with copper wiring.

Arcing inside the electrical box of the weatherproofed outdoor duplex receptacle caused the fire. The arcing most likely resulted from wire insulation that deteriorated due to overheating at one or more of the wire terminals. The wiring at the point of origin was identified as Kaiser KA-FLEX and the receptacle was a Leviton with steel screws. This was the most failure-prone combination that Abbott had been concerned about in his letters to UL.

The CPSC's affidavits and in-depth investigations were part of a multi-faceted program aimed at proving and quantifying the risk and, if warranted, persuading the manufacturers to—voluntarily or otherwise—warn the public of the hazard and pay for corrective action. At the start of the CPSC effort, the agency had funded an NBS project to evaluate the existing field data. The detailed NBS final report, issued in December 1974, sums up the results as follows:

## 12. THE CONSUMER PRODUCT SAFETY COMMISSION

> *The examination of existing field data shows that no available data have the characteristics necessary to develop a reliable estimate of the level of risk to consumers associated with aluminum wiring. Neither can the available data be used to establish the relative risk of aluminum compared to copper wiring. There is only a gross estimate of the extent to which aluminum wiring is now in use in U.S. residences. Statistically sound estimates of risk would be possible only after data collection on a large scale.*

The NBS concluded that the various UL surveys provided only anecdotal data and could not yield accurate statistical estimates. Poorly-worded ambiguous questions and low response rates were serious limitations.

The detailed data that the CPSC sought on wiring methods, failures and fires simply did not exist. If reliable data was required, the agency would have to develop it. The approach that it adopted was to move forward with its own theoretical, field and laboratory work as well as the ongoing hearings, fire investigations and incident report collecting.

A summary of the CPSC's collection of incidents as of early 1975 is contained in the agency's report *Hazard Analysis—Aluminum Wiring*. Its abstract reads:

> *Reports of electrical failures in homes wired with aluminum have become a concern nationwide. According to rough estimates by the Aluminum Association, the total number of homes wired with aluminum in 15 and 20 amp circuits reached 2 million by 1972. The U.S. Consumer Product Safety Commission has collected almost 500 reports of aluminum wire incidents involving electrical malfunctions in single family dwellings, mobile homes, and multifamily dwellings between 1967 and 1975. In these reports, damage ranged from failure of an electrical component to fires with extensive*

*structural damage. Twelve deaths were reported. Summaries of 198 incidents are included in the Appendix.*

W.E. Campbell, Professor Emeritus at Rensselaer Polytechnic Institute (RPI) in Troy, NY, was engaged by the CPSC to develop a theoretical understanding of the aluminum-wired screw terminal failures. Campbell had been a researcher with Bell Laboratories before his retirement. His specialty was surface chemistry, and he was highly regarded as an expert in the field of electrical contact technology.

The CPSC also asked Campbell to propose a laboratory test plan. He recommended against performing the type of high-current "accelerated" test used by UL and the manufacturers. Campbell argued that only tests conducted within the actual conditions of use could genuinely duplicate the failures that were occurring in homes. Using sensitive measurement techniques, connection deterioration with time under normal use conditions could be detected and monitored if and when it occurred. There would then be no argument about whether the results truly reflected the homeowners' experience and risk.

In contrast, UL and the industry relied on a so-called "heat cycle" test, in which current substantially higher than the rated current was turned on and off for defined times for a prescribed number of cycles. UL and industry downplayed failures that occurred in their high current tests, arguing that the test conditions were much more severe than would occur in normal use. Campbell pointed out that the heat cycle test suppressed corrosion related failures that could occur normally in homes and induced failure modes that did not occur in actual installations. He emphasized that environmental factors, such as relative humidity, were likely to be more important than current flow in the failure process.

The CPSC's tests of new aluminum wired receptacles and splicing connectors were performed within actual use conditions, as Campbell had proposed, by Wright-Malta Corp. (W-M) at the Malta Test Station, near

## 12. THE CONSUMER PRODUCT SAFETY COMMISSION  145

Albany, NY. The facility was originally created by the U.S. Army for rocket engine development and was operated by GE until the mid-1960s. A test room was configured to impart daily and seasonal temperature and humidity fluctuations similar to that experienced by the point-of-origin bedroom receptacle in the Hersh home. That receptacle had been in the perimeter wall of the home, with one side of the wall exposed to the outside weather and the other side exposed to room ambient conditions.

Test racks in the room held 1,000 receptacles, 500 pigtail splices made with twist-on connectors and 500 pigtail splices made with a full-compression crimp connector. Half of each group was wired with #12 aluminum, the proper size for 15 amp residential circuits and the other half was wired with #10 aluminum, the appropriate size for 20 amp circuits. The wire, receptacle and twist-on connector samples represented an assortment of brands. The connections in these CPSC tests were subjected to infrequent application of current at less than the circuit rating.

The receptacles were installed in boxes and with cover plates, just as they would have been in a home. Screw terminals were wired and tightened according to the latest industry instructions. The terminals were wired and tightened outside of the boxes and then pushed in to place, just as an electrician would have done to install them in the wall of a home.

Right from the outset, some of the new receptacles ran warm when the current was on, in spite of having been carefully installed according to what the industry promoted as being the proper method. As months passed, the warm-running receptacles became hotter and additional samples started to run warm. Some of the aluminum-wired screw terminals eventually reached the condition of "thermal runaway", which involves rapid deterioration and sudden increase to very high temperature. When that happens, the plastic receptacle body chars and smolders. On those devices, the terminal screws sometimes became red-hot, visible through charred or melted openings in the cover plates.

The red-hot terminal screws observed in the W-M testing demonstrated what people meant when they stated in affidavits that aluminum wired receptacles in their homes were "glowing". One such report, for instance, states:

> ... *June '75—outlet glowing with nothing plugged into it—found completely charred—burned through outlet.* ...

The W-M test data showed that severely overheating—sometimes red-hot—aluminum wired screw terminals generate as much as 40 watts of electrical power at 80% or less of rated current. That is about the same heat generation rate as a kitchen match or the filament of a 40-watt incandescent light bulb. Abnormal heating of one terminal, raising the temperature of the entire receptacle, often accelerated failure of additional terminals on the same receptacle.

The threshold of fire hazard is crossed when an overheating terminal generates about 10 watts. At that heating level the insulation on the attached aluminum wire starts to deteriorate, leading to a potentially hazardous short circuit incident. Fire ignition experiments showed that one screw connection generating 40 watts could ignite a wooden wall stud. That was demonstrated with receptacles installed in plastic or metal outlet boxes, mounted as they normally would be in a wall. The maximum heat generation observed on a single receptacle in the W-M tests was in excess of 100 watts, from simultaneous heating of multiple terminals.

One test at W-M replicated the conditions of the Hersh fire, in which the overheating receptacle terminals were conducting only 4.8 amps current drawn by a combination humidifier/dehumidifier plugged into a receptacle downstream in the circuit. Heat generation of more than 40 watts was measured at one of the overheating receptacle's terminals, demonstrating that fire hazard conditions could develop even at relatively low current.

The hazard posed by aluminum wired receptacle burnouts and burnups in homes was quite obvious, but the manufacturers refused to voluntarily

## 12. THE CONSUMER PRODUCT SAFETY COMMISSION 147

warn the public and pay for corrective actions. The CPSC had to take them to court.

A "quite obvious" hazard demonstrated in a laboratory was not enough to assure that the CPSC would prevail in court. Fire hazards and fires clearly due to aluminum wire connections in homes had to be well documented. And they were.

In 1977, the CPSC conducted an in-home survey of temperature rise in receptacles in Montgomery County, Maryland. Both aluminum and copper wired circuits were tested, with measurements taken after 30 minutes of loading to only 75% of rated circuit current. For homes with receptacles wired at the screw terminals, 18% of the 39 aluminum wired homes had at least one receptacle reaching at least 100 C (212 F), and 41% of them had at least one receptacle overheating to at least 75 C (167 F). None of the screw-connected receptacles in the 57 copper wired homes reached those levels, all being less than 50 C (122 F).

Thirty-six of the overheating aluminum wired receptacles from the homes, with wires intact, were sent to W-M for long-term testing within rated current and environmental conditions. By the end of 1977, three of them had progressed to actual fire ignition in a test wall typical of residential construction.

The CPSC's Montgomery County survey was a pilot project that tested and refined the methods to be used in a nationwide survey. Franklin Research Institute (FRI, later renamed Franklin Research Center) in Philadelphia was contracted to perform the larger survey and the related fire ignition testing. FRI measured receptacle temperature rise in more than 400 homes in four randomly selected areas of the Country: Pinellas County, Florida; Phoenix, Arizona; Baltimore County, Maryland; and King County, Washington. Overheating receptacles discovered in those homes were removed with wiring intact and transferred to the FRI laboratory in Philadelphia. There they were subjected to cyclic loading at rated current.

Twenty-four of them ignited fire in the test wall under conditions of normal use.

The FRI Executive Summary states that 42% of the aluminum wired homes measured contained at least one receptacle that would heat to fire hazard condition under normal use. That was 55 times greater than in the copper wired homes that were surveyed.

At the conclusion of their contract, the remaining FRI specimens that had progressed to fire hazard condition were transferred to W-M, to be continued on test. Additional fire ignitions resulted.

The W-M tests of brand-new aluminum wired receptacles included all types, from steel-screw to newly listed CO/ALR devices with brass terminal screws. Also tested were pigtailing connectors from various manufacturers. These were used with an assortment of brands of aluminum wire, including some listed by UL under its 1971-72 Alloy Assessment Program.

The alloy aluminum wires and the CO/ALR wiring devices were dubbed the *new technology* by the CPSC, in response to UL and industry claims that their latest tests and standards were stringent enough to have resolved the aluminum wire fire safety problem for new installations. CPSC engineers and lawyers were skeptical, but the new alloys and wiring devices had not been on the market and installed long enough in homes to determine if the claims were true. The agency reserved judgment on the question. The CPSC's legal action against the industry only addressed the so-called *old technology*—EC grade aluminum wire and non-Co/ALR wiring devices.

Eventually, more than 6,000 aluminum wire connections were on test at W-M. They were installed properly according to UL and industry recommendations and tested under conditions within the range of normal use. It was one of the largest aluminum wiring test setups in the world, and as far as is known, it was the only one that tested the samples exclusively under conditions representative of proper installation and use. As Campell had predicted, runaway overheating, burnouts, and fire ignition occurred

## 12. THE CONSUMER PRODUCT SAFETY COMMISSION 149

in tests conducted under these normal use conditions, and the connection deterioration process could be tracked from the earliest stages.

The CPSC's test programs at W-M and FRI demonstrated conclusively that old technology aluminum wire and wiring devices posed a unique hazard and that the new technology alloy wire and CO/ALR wiring devices were not failure free. Specifically, they found that:

- Aluminum-wired screw terminals of conventional receptacles, both newly made under laboratory conditions and as they exist in homes, have an abnormally high rate of overheating failure when operating within rated and normal use conditions.

- The overheating failures become more severe with time and can result in fire ignition with receptacles in either steel or plastic outlet boxes, having either plastic or metal cover plates, in walls finished with either sheetrock or wood panel, and with and without combustible material (such as drape or bedding fabric) near the cover plate.

- The action of inserting and withdrawing plugs—which is the basic intended normal use of receptacles—initiated and accelerated overheating failures.

- Most twist-on connectors listed by UL for the purpose had high overheating failure rates when used for the aluminum wire pigtail splice configuration shown in the UL ad hoc committee pamphlet. Those that went to thermal runaway condition became red hot at less than rated current. Two fires ignited in the W-M test racks from failing pigtail splices.

- The only pigtailing splice connector type that survived failure-free in the W-M testing was the COPALUM. This is a full-compression type connector originally developed for aluminum wire used in airplanes.

In August 1975, the CPSC declared that aluminum wire posed an "unreasonable risk" of fire and injury. The manufacturers refused to cooperate. They would not voluntarily warn the public and pay for corrective measures. The agency had to build the case to declare it a hazardous product and obtain a court order to force the manufacturers to act. All of the subsequent CPSC field and laboratory work was in support of that legal action.

In October 1977, the CPSC filed a Civil Action, *CPSC v. Anaconda et al.*, asking the court to declare "old technology" aluminum wiring systems imminently hazardous. Kaiser challenged the CPSC's jurisdiction over the issue, arguing that aluminum wiring was not a consumer product as defined by Consumer Product Safety Act. After a see-saw set of decisions and appeals, Kaiser prevailed. The final appeal decision, issued in January 1982 in favor of Kaiser, declared that aluminum wiring in a home was a consumer product only if the original owner of the home had personally contracted with the electrician for its installation. That was rarely the case, and the decision ended the CPSC's effort to force the manufacturers to warn the public and pay for corrective measures.

The CPSC did retain the right to provide the public with safety warnings and advice as to the hazard and corrective measures that could be taken. CPSC Publication 516, titled "Repairing Aluminum Wiring" was then developed for that purpose. First published in 1984, it contradicted the UL ad hoc committee pamphlet's endorsement of pigtailing with twist-on connectors. The CPSC position, based on test data and field failures, was that pigtailing should be used only for temporary repairs.

Publication 516 superseded UL's ad hoc committee pamphlet in the public domain. The UL pamphlet, originally issued March 1973 and reissued July 1980, was never broadly distributed and has not been available for many years. The CPSC's Publication 516, first issued in 1984, reprinted

## 12. THE CONSUMER PRODUCT SAFETY COMMISSION

in 1994 and 2004, and updated in 2010 is today's authoritative advice to homeowners.

Although the CPSC did not win in court, some of its objectives were accomplished. The agency warned the public, it tested and promoted a repair method that has been failure free and it made available comprehensive data clearly demonstrating the nature and extent of the aluminum wire fire hazard. The CPSC's action and the attendant publicity no doubt weigh heavily as reasons that the market for residential aluminum branch circuit wire collapsed in the mid-1970s and has not recovered. Kaiser's strategy, selling aluminum wire before the connection problems were solved, failed—at great cost to all concerned. It leaves behind a long-lasting fire hazard in several million homes.

The CPSC's branding of aluminum wiring as a hazardous product was based on carefully documented in-home testing, laboratory testing, theoretical analysis, and incident and fire reports from homeowners and fire professionals. The agency's investigations and test results solidly reflected the problems that UL and the industry had long known about and were trying to keep out of the public eye. Aluminum wiring safety problems, well documented, were now on the table for the public to view. This was a major setback for the industry, since the new information contradicted the story it had presented to the public and to government agencies for many years.

The industry's public stance is that lower copper prices killed the residential aluminum wire market. The manufacturers did not contest the fact that the burnouts and fires occurred, but they attributed the failures to "poor workmanship" or "improper installation", blaming the contractors and electricians. UL and the manufacturers claimed, and continue to claim, that aluminum wire is "safe if properly installed".

# 13.
# "IF PROPERLY INSTALLED"

In the opening paragraph of his urgent call for the April 15, 1971 meeting at Travelers Hotel, UL's W.A. Farquhar wrote:

> *Although we believe that present EC aluminum wire, if properly installed, is capable of providing safe service, it is apparent that it is not.*

What does that mean, exactly? Was the aluminum wire not capable of providing safe service, or was it not being "properly installed", or both? The consensus at that meeting was that aluminum wire was not up to the job. In later years, however, Farquhar's words would morph to an industry slogan; "*aluminum wire is safe if properly installed*", implying that if a connection fails, it must have been improperly installed by the person who put it in.

The manufacturers' and UL's concept of proper installation kept changing as the aluminum wire quagmire grew. The resulting confusion is illustrated in a CPSC report:

> *Tenants in twelve apartments ... complained of problems with wall receptacles, switches, and light fixtures. Overheating and arcing were described, as well as a few electrical fires.*
> *... an electrical contractor ... indicated he had found "receptacles*

> backwired with aluminum conductors." This wiring method has never met UL standards for aluminum wire ...
>
> In direct contradiction ... was a statement, dated December 1973, by the wiring inspector of the town of Weymouth, Massachusetts. In his opinion, the original installation was done in accordance with what then prevailed as good design and installation practices with respect to aluminum wiring.
>
> The confusion with respect to specifications for wiring with aluminum in Massachusetts may be representative of similar conditions throughout the United States.

The Weymouth wiring inspector was right. Aluminum wire connected to "backwire" terminals on wiring devices had unquestionably been considered acceptable. UL's original letters of approval left the door open for backwiring with aluminum. Kaiser then explicitly promoted the use of backwiring with aluminum in its 1965 advertising and installation instructions, as in this statement:

> In November, 1964, Underwriters' reaffirmed their 1958 statement that aluminum and copper wires were interchangeable on wiring devices with screw terminals. This includes back wiring as well as side wiring.

The ad says that UL approves of backwiring if the device has both screw and backwire capability. Connecting to a backwire terminal involves pushing the stripped end of the wire into a hole in the back of the device, where it is held either by a spring tab or a screw clamp. That's all there is to it. No time is spent looping the wire to fit under a screw head. Faster installation by backwiring lowers cost for electricians competing for contracts to wire new housing developments.

Following Kaiser's installation advice, with UL's nod of approval,

## 13. "IF PROPERLY INSTALLED" 155

installers across the country legitimately backwired aluminum wire on receptacles and switches and inspectors legitimately granted their approval. Backwiring was not considered "improper installation" in the 1965 time frame. There was no confusion or disagreement at all. There were no objections. The wire and wiring device manufacturers, UL, the electricians and the electrical inspectors were on board at that time, considering backwiring with aluminum as meeting code requirements.

Today, backwired terminations of aluminum wire are universally recognized as being hazardous. It is one of many common installation practices that now, after-the-fact, are frequently called "poor workmanship", wrongly placing blame on the shoulders of the installers and inspectors who simply followed guidance provided by the manufacturers and UL.

In 1965, Kaiser pushed its aluminum house wire into the marketplace faster than UL and the industry were able to respond. The result was a tangled maze of conflicting instructions and opinions on its compatibility with the endless variety of connection types needed to wire a building. UL's chief electrical engineer, Farquhar, tried to clear up the mess with an article published a year later in IAEI News, the electrical inspector association's publication. The piece elicited the following comment from the chief engineer at Bryant Electric:

> *The article in the January, 1966 issue of the IAEI news by Mr. W. A. Farquhar, titled "Aluminum Conductor Termination, A Controversy or a Misunderstanding—Which?" has caused considerable controversy, and, in the writer's opinion, has added confusion to an already misunderstood subject.*

Proceeding at a snail's pace relative to the expanding use of aluminum wire, and still without having done any substantive testing, UL eventually ruled out some of the worst-case terminal types and installation practices. UL's published guidance on receptacle and switch wiring evolved over a decade, sometime changing in contradictory ways as to what devices and which terminals on a device could be used with aluminum, whether or not the wire had to be wrapped around a screw terminal, and if so, how much wrap and in what direction.

The last word from UL and the industry as to connection of aluminum wire to a screw terminal is the version contained in the 1973 ad hoc committee pamphlet. The pamphlet's illustrations have been widely publicized in the half-century since it was first published. Today, anything else is most likely be labeled as an improper installation—poor workmanship—by persons not familiar with the tangle of instructions previously issued.

Up-to-date guidance was slow to reach the people across the country who were actually hooking up aluminum wire in new homes—if it reached them at all. UL had no effective procedure in place to broadcast the information nationwide in a timely manner to those who really needed it. After making a change to its appropriate "guide card", UL notified its clients—the manufacturers—who might pass the information to their salesmen, who might transmit it to their customer contacts, and so on down the various trade and retail supply chains across the country. From time to time, new installation information might be noticed in trade publication articles. Eventually it might reach some of those who were actually installing the aluminum wire in new homes.

Illustrating the lack of clear, uniform and effective nationwide communication on the issue, aluminum wire was still being connected—with inspectors' approval—to backwire receptacle terminals in some areas of the country in 1971. That was about six years after UL modified its "Green Book" statement to rule it out. Ten years after UL declared that backwiring with aluminum was not acceptable, a homeowner in Indianapolis wrote to UL President Whitaker, stating that:

> *I recently had an electrical fire in a receptacle in my house and had an electrician repair damage the next day. As soon as the electrician*

## 13. "IF PROPERLY INSTALLED"

*looked at the damaged receptacle he told me this receptacle was not wired properly, that all aluminum wire must be wired to screw terminals on side of receptacle not stab or quick wired ...*

*I have since talked to two electrical engineers and the head electrical inspector for the state of Indiana, all of which tell me that the stab or quick wire method has never been approved with the use of aluminum wire by the National Electrical Code. But I also talked with the Indianapolis city electrical inspector who says this method of stab or quick wiring with aluminum wire is and always has been acceptable.*

The ongoing use of backwiring with aluminum was supported in part by foggy information from various sources, such as in the "Guide to Use of Devices With Various Types of Conductors" in GE's August 1971 wiring device catalog, which states that:

*Aluminum—May be used only on the terminals of devices which are marked Al-CU on the mounting strap. This applies to all terminals including screw terminals, lug terminals, Pressure-Lock terminals, wire connectors or pressure connectors. ...*

UL added to the confusion by issuing inconsistent and muddled letters, bulletins and articles for trade publications. In March 1972, F.G. McLellan, the Director of Building Safety for Orange County CA, sent a letter directly to W.A. Farquhar, UL Vice President and chief electrical engineer, asking for clarification. In his letter he praised UL's past record on safe and dependable wiring systems, but then added:

*However ... the credibility of your organization is being challenged seriously by the apparent misunderstanding as to what consists an acceptable method of termination of ... aluminum conductors ... in residential branch circuit wiring. Based on your recommendations, this department took a positive stand which we recently have had to reverse. So many conflicting statements are being accredited to your organization that I am taking the liberty of requesting some information to assist me in understanding the position you are taking on aluminum conductor termination.*

Five years later, in 1977, UL and the industry still did not have wiring instructions clearly sorted out. UL engineer R.F. Gloisten wrote as follows regarding comments by an electrical contractor:

*… Mr. Pernick is concerned that the installer of this receptacle may not know which wire to use where. As you can see in my letter (to him), I sympathize with his frustration in trying to get a simple and a specific indication of what wires should be used. …*

Regarding that same complaint, UL's Supervising Inspector in Los Angeles stated that:

*There is no doubt that the marking is so confusing that the installer and inspector will not understand it and errors will occur.*

Proper installation can logically be defined as following the instructions that come with the product. Or, if no instructions are provided, following generally accepted practice. Manufacturers of house wiring components were not required to provide instructions for connecting wire to screw terminals. Long-established wiring practices worked well with copper wire. During the early years of its use for house wiring, aluminum wire manufacturers encouraged the notion that its installation was the same as copper wire except for selecting the next larger size for a given circuit current rating,

## 13. "IF PROPERLY INSTALLED"

casting aside well-founded requirements for use of corrosion inhibitor and abrasion for aluminum connections.

Until Kaiser introduced its KA-FLEX aluminum wire for house wiring, the accepted practice for connecting aluminum included application of a corrosion-inhibiting grease compound and abrading the wire surface to remove the insulating aluminum oxide layer. Although not a cure-all for the aluminum connection problems that were occurring, the use of inhibitor and abrasion substantially reduced the risk of runaway overheating failures. Kaiser's first instruction manual for interior wiring, in 1954, states that:

> *Penetrox* [an inhibitor compound] *should be liberally applied to aluminum conductors. After application, the conductor should be rubbed with an abrasive cloth or scratched through the compound with a wire brush.*
>
> *In some areas where there are no corrosive influences, it may be possible to make satisfactory aluminum connections without the use of any inhibitor; however, this is difficult to determine, and connections may deteriorate under supposedly favorable conditions— therefore never, UNDER ANY CIRCUMSTANCES MAKE UP PRESSURE CONTACT CONNECTIONS WITHOUT AN INHIBITOR.*

The capitalized emphasis is Kaiser's. But the directive to use inhibitor and abrasion worked against the company's objective to sell aluminum. Extra installation time was required to apply inhibitor and abrade the wire, and that would reduce or eliminate aluminum's cost advantage relative to copper. Kaiser's justification for eliminating inhibitor and abrasion from its installation instructions in 1964 was that UL did not use it when testing connectors with aluminum conductor. UL, on the other hand, tested that way to represent the "worst case" installation when a workman doesn't follow installation instructions or accepted practice regarding inhibitor and abrasion.

Kaiser went a step further, not only eliminating it from their installation instructions but actively promoting the notion that inhibitor and abrasion were not required. As an example, instructions that came with 3M "Scotchlock" twist-on connectors stated that, for aluminum wire, an inhibitor compound should be applied "according to the manufacturer's instructions". Kaiser salesmen were directed in 1965 to tell customers that this referred to the wire manufacturer's instructions, not those of the inhibitor compound manufacturer, and that Kaiser's instructions were that inhibitor compound was not required. That was not based on any testing, and it was at odds with the company's own experience at its trial installation at Ravenswood. There, a Kaiser engineer reported Scotchlock twist-on connector "problems" that he attributed to installation without inhibitor.

In the price-competitive market to supply wire for new housing, instructions and cautions regarding use of the product could put a damper on sales. Companies issuing restrictive or labor-intensive instructions would be at a disadvantage. Reviewing whether his company included instruction sheets ("stuffers") packaged with the wire, Kaiser's Roger Keith wrote in late 1972:

> *I don't have any sure way of checking instruction sheet activity during late 1967 but am reasonably certain from early 1968 through June 1971 we did not use instruction sheet stuffers as competition was very active beginning in 1968 and none use stuffers.*

For at least three years, therefore, while aluminum wire was being installed in about two million new houses and mobile homes, the manufacturers did not include installation instructions with their product. Considering this, in combination with UL's confusing guidance, it is not surprising that installations with so-called "poor workmanship", "improper installation" or "improper wiring devices" can be found in the majority of the

## 13. "IF PROPERLY INSTALLED"

aluminum wired homes. When actual samples of burned aluminum-wired connections were examined by UL and industry personnel, the analysis seldom went beyond a simple visual inspection, identifying backwiring, too little or too much wrap around a terminal screw, or wire wrapped in the "wrong" direction so the failure could be blamed on the installer.

It came as a surprise to many in the industry when UL started to call counterclockwise wrap of the wire at a terminal screw "poor workmanship". Most UL listed receptacles commonly available at the time were designed in a mirror image fashion that encouraged and sometimes forced installation with one clockwise and one counterclockwise wrap on each side. The proposed UL test procedure for aluminum-rated CO/ALR wiring devices, in 1972, recognized this, specifying that the wires should be wrapped counterclockwise if dictated by the design of the terminal.

If a failed terminal did not have any visual indication of "improper" installation, the overheating could still be blamed on the installer by implying that it must not have been sufficiently tightened. That is about the only cause of connection problems with copper-wired screw terminals. Ignoring the many documented reasons that can cause an initially tight aluminum wire connection to loosen, it is easy for someone so inclined to conclude that the initial installer did not tighten it enough. Whether or not it was initially well tightened, an aluminum-wired screw terminal might not feel tight when checked with a screw driver.

In 1971, UL's O.G. Wedekind listed several design characteristics of the aluminum wired receptacles that led to their overheating in his test. He did not identify poor workmanship as one of the contributing factors. The receptacles had been installed by Wedekind himself and/or a UL technician.

Wedekind had also inspected some failed aluminum-wired devices from homes in his area. A majority of these were kitchen receptacles. His analysis led to the following conclusion as to one possible cause of the overheating connections:

> *... the design of the conventional 15 ampere receptacle having wire binding screw terminals makes it very difficult for the electrician to make an effective connection with No. 10 aluminum conductors since these terminals were never really designed for this size of wire. The head of the terminal screw is too small and the space between the screw head and the insulating walls of the terminal cavity make it difficult to wrap a wire around the terminal screw even when the screw is backed out to the limit ... the wire very often fails to be forced against the terminal plate by the terminal screws ...*

Notwithstanding such forthright analysis from within UL itself, the wording of Farquhar's call to the April 15, 1971 meeting reflects a turning point in UL policy. Prior to 1971, UL communications that touched on the causes of the aluminum connection problems generally indicated that they were due to many contributing factors, some still unknown and most of which were out of the control of the installer. Subsequent to 1971, UL communications on the matter imply one way or another that the original installers are to blame for the burnouts and fires.

Projecting the poor workmanship argument to the world outside of UL, Farquhar generally relied on anecdotal hearsay and opinions from contractors and inspectors who supported the use of aluminum wiring in homes. UL never published or provided substantiating reports, analysis or data. Farquhar essentially assumed the role of defense attorney, picking and choosing only evidence favorable to UL and its clients to submit to the jury. He essentially dusted under the rug any and all research reports, failure analysis and test data from UL or outside sources that cast doubt on the UL poor workmanship conclusion.

The poor workmanship argument also touched the so-called "new technology" aluminum wire and its connections. Farquhar claimed that the reason they were developed was to be more tolerant of poor workmanship—to make them "more foolproof". Taking it to an extreme, UL

## 13. "IF PROPERLY INSTALLED" 163

disregarded early failures in the qualification testing of the "new technology" aluminum wire alloys, based on the assumption that early failures must be the result of defective installation by the UL or manufacturer's personnel who wired the test specimens. When one of Leviton's new technology ("CO/ALR") receptacles failed in its initial testing at UL, the failure was blamed on insufficient tightening of the terminal(s) and was not counted against approval of the product.

A persuasive refutation of UL's poor workmanship argument for overheating failures of aluminum wire terminations and splices comes from within UL itself. In 1977 Wedekind updated Farquhar on the results of his long-term test of ten aluminum wired receptacles. The connections were tested with an on/off cycle that never exceeded the rated current of the circuit. His test wall circuit incorporated some wire splices made with UL listed twist-on connectors, generally known as "wire nuts". Wedekind noted that the receptacle temperatures seemed stable (after having retightened terminal screws on several that initially overheated), but added:

> ... in fact I have had more difficulty with the wire nut splicing connectors ... several of them which [sic] have experienced runaway conditions even though I took special care to assure a good connection. ... burned up several ...

As UL narrowed its focus in 1971 to blame the installers, results of various research and testing projects initiated only a year or two earlier pointed to the intrinsic properties of aluminum as being at the root of the problem. Kaiser's own people came to understand that "poor workmanship" was not the primary cause of the aluminum wire connection burn-outs. In a mid-1971 memo to T.R. Pritchett, head of the company's research lab, the Center for Technology, Kaiser researcher S.G. Roberts wrote:

> *The poorer "connectability" of aluminum relative to that of copper*

> *appears to be basically due to: aluminum's highly-insulating and rapidly forming oxide, its relatively high thermal expansion and the low creep-resistance at elevated temperatures of EC grade aluminum. These characteristics combine to provide a mutually accelerating degradation of the electrical contact in a connection. ...*

At the same time, R.L. Horstman, President of Rodale, a wiring device manufacturer, wrote to Congressman F. Potter citing the intrinsic properties of aluminum as the underlying cause of the "serious safety hazard", and did not even mention "workmanship" as a factor. He wrote:

> *As I indicated to you briefly on the phone last week, the continuing use of aluminum wire for branch circuit wiring (residential) represents a serious safety hazard which must be researched and controlled.*
>
> *Aluminum possesses two unfortunate and dangerous characteristics which evidence themselves when applied to ... switches and receptacles. Even the newly researched alloys that are being developed by the aluminum companies haven't been able to overcome properties of potential oxidation and cold flow ...*

Despite strong evidence to the contrary, Farquhar's strategy gained traction. Blaming the installers became the principal defense strategy of UL and the manufacturers in their efforts to dodge liability for the enduring safety problems in aluminum wired homes.

The manufacturers eventually closed ranks and climbed aboard UL's poor workmanship bandwagon, most likely as the legal departments overshadowed the engineers in guiding company policy. The focus shifted to limiting the companies' financial liability rather than solving the customers' problems. The lawyers are the "elephants in the room" in this story, since documentation of their activity is not available through the discovery process

## 13. "IF PROPERLY INSTALLED"

in product liability litigation. In many instances throughout this book, the actions of people and companies might only be understood by recognizing the lawyers' unreported influence.

Leviton, a major wiring device manufacturer, did not accept UL's poor workmanship story for the first few years after it was rolled out. In mid-1971, M.J. Weitzman, Leviton's VP-Engineering commented on a meeting report from UL which stated that examination of failed and newly made aluminum-wired screw terminals:

> ... *appears to indicate that the problems result from poor workmanship—inadequate tightening, failure to wrap the conductor around the screw in the proper manner etc. ...*

Weitzman countered that UL should face the problem squarely and own up to the fact that copper wire works fine under the same installation conditions and that the underlying cause of the failures is the aluminum oxide on the surface of the wire.

Leviton's patent for its improved terminal for aluminum wire, applied for in 1974, discusses causes of the failures and explains how the invention deals with them. Poor installation workmanship is not mentioned as one of the causes.

By 1978, the company's posture had aligned with UL's. Responding to questions posed by Congressman Moss, Leviton's J. Pearse (Weitzman's successor) parroted the UL poor workmanship script without any discussion of the problematic properties of aluminum. Pearse referred to UL actions and conclusions, but did not provide any of his own company's analysis and test results. (On his copy of Pearse's letter to Moss, Alan Schoem, an attorney with the CPSC, wrote "what a crock", the same sort of comment that Weitzman and company President H. Leviton had written from time to time after reading UL communications on the subject.)

The National Fire Protection Association (NFPA) also climbed aboard the "poor workmanship" bandwagon to defend the National Electrical Code that it publishes. For example, in 1991, an electrical specialist with NFPA responded to an inquiry from the owner of an aluminum wired condo unit with the following re-write of history:

> *Aluminum wiring is permitted by the National Electrical Code. When this product came on the market, the installation methods were quite different from than [sic] traditional procedures for installing copper wire. The special instructions provided by the manufacturers were often times overlooked or not thoroughly followed. The result was that certain connections were likely to fail after a period of time.*

The NFPA electrical specialist who wrote the above response likely relied on a story that was provided by UL and industry and massaged by the lawyers. First-hand knowledge and experience gained a half century ago has largely disappeared. Those who originally participated in the activities and meetings have retired and/or passed away. UL and industry research and test reports have been tightly guarded and never publicly released. CPSC documentation of its aluminum wire investigation is generally available only through freedom of information requests. UL's poor workmanship story is often repeated but seldom, if ever, vetted.

Blaming the installers is a double-edged sword for the industry. Electricians may be reluctant to use aluminum wire if they believe that the manufacturers will turn against them again if problems arise. It also proclaims that aluminum wire cannot be connected properly using the same workers, skills, and methods used successfully for a century to connect copper wire. These factors, along with the lingering bad reputation of aluminum wire, serve to inhibit its return to the market no matter what its price advantage may be over copper.

In advertising, promotional literature and trade publication articles, some manufacturers push to overcome the resistance and pave the way for

## 13. "IF PROPERLY INSTALLED"

a revival of the aluminum NM cable market in home construction. The poor workmanship argument is repeated time and again while other causative factors are whitewashed, misrepresented or just ignored.

A 2006 article in IAEI News (the electrical inspector association's magazine) serves as an example. Written by an ALCAN field engineer and titled, "Still Living in the 60's? Aluminum Building Wire and Terminations", one-third of the eight-page article addresses branch circuit house wiring. The installers are blamed for the pre-1972 problems, citing the usual factors already discussed and the use of wiring devices "meant only for copper".

Of course the installers used receptacles and switches "meant only for copper". That was what the industry told them to do and, until 1973, they were the only ones available. They were produced with and without the "Al-CU" marking, but that symbol did not reflect any difference in design, materials, performance, or qualification testing relative to the same brand and model without the symbol. The AL-CU symbol on wiring devices was, as Leviton VP-Engineering M.Weitzman wrote, "a marketing gimmick". Devices with the symbol were distributed to regions where local inspectors insisted on it. There was no choice but to use "devices meant only for copper" until 1973, when the first "CO/ALR" devices, specifically tested for use with aluminum wire, became generally available. The author of the 2006 article in IAEI News deserves a little bit of sympathy for misrepresenting the installer's situation. She did not have first-hand knowledge of these details. She was a little girl when it all happened, and her industry sources were not likely to have disclosed that information to her.

A dictionary definition of "workmanship" is *"the skill with which something was made or done"*. Applied to UL and industry actions, perhaps "poor workmanship" on their part truly underlies the aluminum wiring fire safety problem. How else to characterize UL's recognition of potential fire safety problems as early as 1946 and then waiting until 1971 before starting substantive testing? How else to characterize Kaiser's strategy of introducing and selling aluminum building wire before its connection problems were solved?

Poor workmanship also surfaced at the aluminum wire factories from time to time. Troublesome product quality issues at Kaiser resulted in some defective KA-FLEX NM cable reaching the marketplace and being installed in homes. In the worst case, the defects could cause fire ignition.

# 14.
# FATAL FIRE—PHOENIX ARIZONA

The Erskine family was enjoying dinner at a friend's house on a hot day in May 1987. Brik Erskine walked back home, a block or so away, to fetch something, but didn't return. Investigators surmised that he discovered the fire upon entering the house and then was overcome by smoke while doing whatever he tried to do next. He left behind his wife and two small children.

It was determined that the fire originated in the attic space, but the exact cause could not be pinpointed. The area of origin encompassed mainly some Kaiser KA-FLEX aluminum NM electrical cable runs. For the most part, the cables in the area of origin were no longer there. Plastic insulation and cable jacket material burned away and the aluminum wire melted. No informative physical evidence remained at the area of origin. The cable sections leading to and from the area of origin showed no signs of abnormal high current.

Plausible sources of ignition included sparks from a nearby electrical box or some sort of flaw in one of the cable runs going through the area of origin. Ignition at a point along a run of cable at normal operating current is rare, however. It would not be expected to happen on a run of typical NM cable in a house unless there is some sort of physical damage. But the cables in the Erskine home may not have been typical.

Wire breakage was a thorny problem in Kaiser's KA-FLEX manufacturing lines. The wire itself was softer and more flexible than previous offerings of aluminum building wire. That was a key selling point. But its lower tensile strength made it more likely to break during manufacturing operations as it progressed from machine to machine and was wound on or off reels.

Welded "process joints" were commonly employed to connect wire ends when starting a new reel of stock or recovering from a stoppage caused by a break. Process joints cannot be relied on to provide the same properties as the wire itself. They can fracture or break from stresses imposed during additional manufacturing steps or during installation in a building. In the production of stranded (multi-strand) building wire, occasional welded process joints occurring at random on individual strands can be left in the finished product without any adverse consequence.

In contrast, process joints remaining in single-strand (solid) conductor cannot be tolerated, since they compromise both reliability and safety. They may fracture, causing the circuit to be open, intermittent, or a develop a hot spot at that point of the cable run. Removal of all process joints is generally required for finished NM cable shipped from the factory.

Branch circuit NM cable, such as Kaiser's KA-FLEX, is typically constructed with three or more solid wires, each of which is insulated except for the bare "ground" wire. Sections of the finished cable that have process joints in them must be cut out and scrapped.

Not all process splices were welded. A length of aluminum NM cable with an abnormal bulge was examined by the CPSC. Radiographic (X-Ray) imaging of the bulge revealed that one of the wires had been spliced by simply twisting the wire ends together. In an active circuit in a home, that sort of splice along a cable run poses a serious fire hazard. The defect would not have been detected by the continuity test normally applied at the factory before shipping.

A continuity test determines if a wire in the cable conducts electricity from one end to the other. The most reliable type of continuity test involves touching instrument probes to a wire at the two ends of the coil, one wire

## 14. FATAL FIRE—PHOENIX ARIZONA

at a time according to the color of the wire's insulation. Black and black, white and white, and so on. The instrument rings a bell, lights a light or sounds a buzzer if current properly flows from end to end. There should be zero failures for this test if the manufacturing process is well controlled. Kaiser applied the "bell" test to some sample coils out of each lot shipped as a quality control measure, but this still allowed a significant amount of cable with process joints and other defects to enter the market supply chain.

One remarkable quality control report for a lot of 1,000 coils of KA-FLEX noted seven continuity test failures in a sample of ten coils tested. The inspector entered 99.3% as the quality level and indicated that the lot, minus the seven coils that failed the test, was OK to ship. That shipment alone, with 70% failure rate for the samples that were tested, most likely contained hundreds of defective coils.

That was not necessarily a big problem for Kaiser. If a visually-obvious defect was spotted before the cable was installed, it might only take a refund for the cost of the cable and/or few kind words to smooth things over. In 1967, for instance, a contractor in Colorado noticed an obvious defect and sent that section of cable to Kaiser. The complaint form describes the defect and the outcome as follows:

> *... Looks like one of our operators had a bad night ...*
> *... Partway through the 250' coil the material was noted to be broken and was sort of held together by plastic tape on one conductor with the conductor twisted around itself. The particular section in question was returned to Tom Brown at Bristol some 6 weeks ago ... Tom wanted an informational claim to show as a record.*
> *No dollars involved as Carl bought Louie and the contractor lunch ...*
> *Try not to let material get out of the mill in this condition.*

If there was no visual clue, electricians might discover defects only after the cable was installed within finished walls and ceilings. Circuits would become intermittent or would not work at all. They had to diagnose the problem, determine where the break was, and replace that particular cable run. Even then, installers finding a defect now and again were not likely to do anything more than just replace the defective section. An electrician who took the time to follow through with a complaint would get a refund for the cost of the cable, and perhaps some compensation for time spent tracing down the problem and replacing the defective cable section.

Arthur Electric in Phoenix, Arizona, where the Erskine fire occurred, encountered multiple instances of breaks, splices and missing wire insulation in KA-FLEX over several years of wiring homes with aluminum. In August of 1968, for instance, the company found defects in two different lengths of cable, one having an uninsulated section of one wire with a process weld in it and the other having a taped-over "U splice" in it. In April of 1973, Arthur Electric processed a claim involving five instances of KA-FLEX with discontinuous conductors.

Inevitably, not all of the defects were caught by the installers. Some homeowners and occupants were left with lingering hazards in their wiring. A current-carrying splice or flaw of any sort in mid-run of a cable is as hazardous as an improper connection anywhere else in the system. Some cable defects came to light when a circuit simply stopped operating. An affidavit submitted to the CPSC by the original owner of an aluminum-wired mobile home, for example, states that four years after the home was purchased:

> *Discovered that refrigerator was not working. After tracing trunk line and performing continuity checks it was determined that the line had failed somewhere between the cross over point and the outlet. A new line was installed.*

Others have not been as lucky. Fires originating at a point along a run of aluminum NM cable have occurred. San Diego, California is another place where in-line defects in KA-FLEX cable were discovered by installers.

## 14. FATAL FIRE—PHOENIX ARIZONA

Three years later, in that same city, two children died in a fire determined to have started at a point along a run of aluminum NM cable. The city's chief electrical inspector did not understand how that could have happened.

In Thousand Oaks, California, a fire occurred in the attic of an 11-year old aluminum-wired house, closely resembling the situation in the Erskine fire. The CPSC's in-depth investigation report summarizes the incident as follows:

> *A fire occurred in the crawl space of an attic, on 3/10/79, apparently caused by a short in a Romex type length of aluminum wire. The point of ignition appeared to be at a point on the wire where it was suspended in air; fire appeared to be electrically ignited inside the cable sheath. Fire damage was mostly structural to the attic and roof ... No one was injured.*

Kaiser KA-FLEX NM cable with conductor and insulation defects did ship to Phoenix multiple times during the years that it was being installed in homes such as the Erskine's. The type of defects noted have the potential to cause fire ignition. That conclusion, based on electrical safety fundamentals, has been bolstered by actual incident reports. But the exact cause of the fire that killed Brik Erskine could never be determined with absolute certainty.

A product liability lawsuit on behalf of his surviving family went to trial. It was a "bench" trial, without a jury. Kaiser settled part way through plaintiff's presentation of their case, apparently coming to believe that the trial judge was likely to agree that their product most likely caused the fire.

# 15.
# CANADA

The Toronto Housing Authority noted that seven fires in its properties were caused by aluminum wiring in 1975, "... about the same as last year."

Apparently living in a different universe, the head of the inspection department in Ontario, at about the same time, addressed a meeting of concerned homeowners in Scarboro, one of the boroughs of the city of Toronto. A newspaper article reported as follows:

> *If your house has aluminum wiring, stop worrying. It won't cause a fire, an Ontario Hydro official insists. And furthermore, there has never been a fire in Ontario or in Canada, for that matter, that could be directly attributed to aluminum wiring.*
>
> *Jack A. Dicker, manager of the inspection department of Ontario Hydro, told 100 Scarborough residents at a special public meeting ... that extensive investigation of aluminum wiring had shown that it is safe. ... Facts support that there is no fire hazard resulting from the use of aluminum wire.*
>
> *... investigators found that there was a higher than expected incidence of receptacle overheating failures with aluminum wire than with copper wire, "but the damage was confined to the metal box in which the receptacle is contained and no fire resulted," he said.*

*... In many cases where there have been failures, the receptacles had been installed improperly, he said.*

*Both Scarborough Controller Joyce Trimmer and M. Walker Broley, chairman of the Scarborough Public Utilities Commission have had receptacles melt in their homes, which have aluminum wiring. ...*

Joyce Trimmer's position as Controller of a Toronto borough was similar to that of a Borough President in New York City. She had been drawn into politics years before, as an activist on neighborhood issues. As Controller, she was popular and well respected. The aluminum-wired receptacle incident in the family's own house brought her into the camp of homeowner activists who were dealing with the same hazard.

She was outside of the house and saw smoke coming from the open window of a second-floor bedroom. The smoke was issuing from the window curtain, which hung down across the face of a receptacle. No one was in the bedroom at the time. The receptacle that charred the curtain overheated from current passing through its terminals to a hair dryer being used in another room. Quick response limited the damage, but the prospect of fire ignition from failing connections in the aluminum-wired home was suddenly quite real for the Trimmer family.

Many other families shared that concern after incidents in their own homes. The homeowners bottom line was that aluminum wiring is not safe. Frequent hazardous overheating at connections would not be happening if their homes had been wired with copper. Family safety was their primary objective.

Several homeowner organizations formed. They urged authorities to halt the use of aluminum wire and initiate corrective action for existing installations. Gary Heighington, co-founder of one such group, presented their cause at meetings and hearings. He underscored the hazard that the homeowners were dealing with by showing a display board crowded with

## 15. CANADA

charred aluminum-wired devices from homes. The homeowner groups faced a wall of opposition from ALCAN, CEA, CSA and Ontario Hydro.

ALCAN produced the aluminum metal and also manufactured wire and cable, just as Kaiser did south of the Canadian border. As a major employer, ALCAN was influential on political and economic fronts in Ontario Province and across the country. Its denial of hazard and opposition to bans was no surprise, since both sales and potential liability were at stake.

CEA (Canadian Electrical Association) would naturally support its members and take the same position. It is an industry advocate for both manufacturers and electric utilities.

CSA (Canadian Standards Association) is often thought of as the northern twin of UL. But there are substantive differences in overall scope of activity and responsibility, as explained (1977) by the chief engineer at Bryant:

> [In the USA] *UL standards are written entirely by UL, with the manufacturers' representatives utilized in an advisory capacity only ... The manufacturers can wield considerable influence, but UL retains the ultimate responsibility. UL also has an Electrical Counsel which they use as a sounding board for changes. This group is composed largely of inspectors.*
>
> *The National Electrical Code is sponsored by NFPA and is made by committees composed of representatives of manufacturers, UL, insurance companies, end users, contractors, government, and inspectors. This is an unofficial document giving recommendations and is official only upon adoption by the government of a city or state.*
>
> [In Canada] *CSA Standards ... are written by Spec Committees ... comprised of representatives of CSA, the manufacturers, and the inspection authorities. Theoretically, the manufacturers have greater say than in the UL process but in actual practice, it is hard to know whether this is correct. The inspectors do have much more say, with*

> *the ones representing Ontario Hydro having the most say, probably because of geographical location and because of the percentage of population they represent. ...*

Ontario Hydro was the government-owned electric power utility in Ontario until going private in 1999. It was responsible for power generation and distribution, electrical inspections and code enforcement, and it conducted some laboratory research. The electrical inspectors in Ontario worked for Ontario Hydro. They spoke with one voice on aluminum wire issues, as directed by their employer.

Throughout years of controversy on this topic, in both the U.S. and Canada, there were no common concepts of what "fire", "fire hazard", or "caused by" meant in the context of the aluminum wiring problem. Is the old saying, "where there is smoke, there is fire" good enough? Is a smoldering fire (without flame) a *fire*? Can an incident be called a *fire* if the fire department is not called? Is it a *fire* if the damage is limited and no insurance claim is filed? Was the incident at the Trimmer home in Toronto a *fire*? Was it even proof of a *fire hazard*? Was it *caused* by the aluminum wire, or could it be blamed on use of an improper wiring device or poor workmanship? The industry, CSA, and Ontario Hydro exploited these ambiguities of language and concept in their denials, trying to dodge blame and liability for the homeowners' plight.

CSA, for example, pulled out all the stops in a 1975 article in its periodical flyer, "Standards Canada". The article characterized the connection overheating problems as nuisances, stated that CSA and Ontario Hydro did not consider them to be hazards, claimed that there was nothing wrong with the receptacles that were used, and blamed the problems on poor workmanship. Regarding fires, the article said:

> *The Approvals Council (Electrical), which consists of provincial electrical inspectors from across the Canada, have stated that there*

*has never been a fire in Canada that can be directly attributed to the fact that aluminum wire was used.*

Of course not! No coil or piece of aluminum wire just flamed up all by itself. And there never was an overheating problem unless the aluminum wire was attached by an installer (who could be blamed) to some sort of connecting means (that also could be blamed). Delete the word "directly" and the CSA statement becomes provably false.

The CSA article also stated that, *"Aluminum wiring has been used in Canadian homes for the last 10 years."* That was a stretch. Aside from 75 houses in Arvida, Quebec, very little aluminum wire was used in Canadian homes prior to 1970. Arvida originated as an ALCAN company town. Its name derived from the name of the company's founder, Arthur Vining Davis. The aluminum-wired houses in Arvida, constructed in about 1948, served as a pilot installation, analogous to Kaiser's at Ravenswood. It has been touted by ALCAN and Ontario Hydro as having been problem-free.

In 1970, when aluminum wire fire safety problems were catching the public eye south of the border, installation of aluminum wiring in Canadian housing was just starting to pick up steam. A letter to UL from Canada Wire and Cable, in 1970, reads:

*I notice you held a meeting on this subject on February 20th. The introductory paragraph states you are receiving continuing reports of field failures of connections.*

*As we are just beginning to use aluminum conductor building wire in Canada we are deeply interested in this activity. Can you provide me with some detail regarding the type of field failures you have encountered? ...*

The delayed introduction of aluminum wire for home construction in Canada in part reflects caution due to previous experience. Canadian responses to one of the UL surveys indicate that connection problems had been experienced with aluminum building wire in prior years.

In 1968, three years after Kaiser introduced KA-FLEX in the U.S., aluminum branch circuit wiring started to be used in Canadian home construction. ALCAN asked UL for information on how it tested aluminum branch circuit connections. The engineer assigned to respond was instructed to reply by sending a copy of UL's Bulletin of Research #48. But that 1954 report does not answer the question. It simply documents experiments intended to point the way to an effective test method for aluminum connections. UL had little else to offer though, since it had not yet established test procedures and had no relevant data.

Subsequently, in 1969, the door opened a bit for Canadian representation at some of the UL and industry meetings on the use of aluminum wire. The extent to which the field failures in the U.S. were disclosed and discussed at meetings attended by the Canadians at the time is not known. Whether Canadian observers at any such meetings had an obligation to educate the CSA and their country's electrical industry about the Yankees' aluminum wire difficulties is not clear.

It is clear, however, that the sense of what was happening in the States did not get through to key people at CSA in a timely manner. In late 1970, at a NEMA meeting, CSA's D.M. Manson was reported as being "jarred" when he learned the extent of the Americans' aluminum wire problems. Manson was head of CSA's Standards Engineering Department, and had just previously signed the CSA Bulletin allowing push-in backwire terminals to be used with aluminum wire.

CSA's acceptance of push-in (backwired) terminals for aluminum in 1970 provides a vivid illustration of a certification process gone awry, not unlike that which precipitated the problems in the States. The approval flew in the face of many objections. At the NEMA meeting, Manson indicated that the opposition, from many parties, was too general and not backed up with test data.

## 15. CANADA

One of the objections was from P. Haskelson, Manager of Leviton's Canadian subsidiary, who wrote to CSA:

> *... I believe that in five years we may be sorry to have allowed the use of aluminum wire with binding head terminal screws but if we now allow the use of push-in with aluminum wire we will be guilty of criminal negligence.*

CSA's backwiring gamble failed, and homeowners were the losers. Shortly after CSA opened the door of opportunity, Canadian manufacturer Smith and Stone (S&S) introduced backwire-only wiring devices having the required AL-CU marking for aluminum. Costing less to buy and install than devices with screw terminals, they were used in thousands of new dwellings in the "survival of the cheapest" competitive market. In less than four years, after many of them overheated in service, CSA revoked the certification. These receptacles remain installed in many Canadian homes and hazardous incidents of overheating continue to occur.

This stands as a strong rebuttal to the poor workmanship excuse for overheating aluminum connections. There was no way other than push-in backwiring for the installer to connect the aluminum wire to these CSA-certified-for-aluminum receptacles. In a sense, it reflects poor workmanship on the part of CSA and S&S, not the installer. Underlying this failed experiment is the fact that the industry did not have a testing method stringent enough to reliably separate connection designs that would perform properly from those that would not in aluminum-wired homes.

Burnouts of twist-on splicing connectors ("wire nuts") with aluminum wire were also systemic in Canada, providing another demonstration of inadequate standards and qualification testing. Electric heating is much more common in Canada than in the States. With the introduction of aluminum wiring came reports of burned and red-hot splice connections in electric heating equipment.

CSA responded by issuing a more stringent standard for a so-called "special service" connector. When these new connectors became available, they were independently tested by Wright-Malta Corp. (W-M), south of the border. Operating in normal environment and at less than rated current, test splices started overheating within a year. The "special service" connectors on the market proved to be failure-prone in the intended application—connecting Canadian solid aluminum wire to stranded copper wire. The new CSA standard had not served well towards solving the problem.

Professor Bernard Béland at the University of Sherbrooke in Québec also tested Canadian aluminum-wire splices made with CSA-certified twist-on connectors. Béland was well-regarded for his promotion of accurate fire investigation. He rightfully campaigned against the common practice of blaming a fire on electrical ignition if other explanations were not obvious. Béland had faith in CSA and UL. He was skeptical that connectors certified and listed by those well-regarded organizations could be failure-prone. At an electrical fire safety seminar in San Diego, he was challenged to set up a test of aluminum-wired twist-on connectors and see for himself. That he did.

Béland was engaged as an expert witness for defendants in the Beverly Hills Supper Club aluminum wire product liability lawsuit. The lab notes and data for his twist-on connector test, in French, were among the items provided to plaintiffs' attorneys prior to his scheduled deposition. Shortly after a translation of Béland's lab notes was made available, defendants removed him from their list of trial witnesses.

The lab notes showed that a substantial number of his test splices had overheated by the end of a year of operation within rated conditions. Plastic insulation melted, charred, and burned. Some splices became red hot. Having obtained results similar to W-M's, Béland nevertheless did not fault CSA, the aluminum wire or the connectors. Because only some specimens—but not all—in each test group overheated, he took the position that the ones that failed must have been improperly installed by his lab technician.

# 15. CANADA

Following the alloy development path taken by aluminum wire producers in the U.S., ALCAN released a new aluminum alloy wire under the brand name NUAL. This new Canadian NM cable was also evaluated at W-M. Overheating failures among the test connections developed once again. So far, no combination of new connectors or new alloys solved the problem. Nor could it for aluminum-wired homes that had already been built.

The homeowners' view, that aluminum wire in the homes posed a unique hazard, unlike anything experienced with copper wire, gathered momentum and gained press coverage. In April 1977, the Ontario provincial government responded, establishing the Commission of Inquiry on Aluminum Wiring. The Commission was charged with investigating, holding public hearings, and making recommendations on the reliability and safety of aluminum residential wiring relative to copper wiring. It is generally referred to as the "Wilson Commission", named after its appointed chairman, John Tuzo Wilson, a geophysicist. Wilson is recognized for his contributions to plate tectonic and continental drift theories. A mountain in Antarctica and a dormant undersea volcano off the west coast of Canada are named in his honor.

The Wilson Commission submitted its report at the end of September 1978, 18 months after its formation. The Commission had conducted hearings, ingested submitted material, and reviewed the status and problems of aluminum wiring in Canada and some other countries. The report is lengthy, but most of the content is only remotely related to the assigned task—assessing the relative reliability and safety of aluminum vs. copper wiring *in homes*. The few statements that address that objective head-on are needles of information in haystacks of irrelevant padding.

Most of more than 100 pages of padding is from testimony and material provided by ALCAN, CSA, and Ontario Hydro, the principal sources of "expert" testimony recognized by the Commission in its court-like hearings. There is a tutorial on the residential electrical system, eight pages long and illustrated with more than a dozen photographs. There is an incohesive hodgepodge of technical information on electrical contact theory, a disjointed collection of experimental results, an extensive history of aluminum wiring, and discussion of the standards and actions taken as problems with aluminum connections erupted.

In contrast, the homeowners' actual experience with aluminum connection burnouts and fires is barely mentioned in the Wilson Commission's report. The in-person testimony of thirty homeowners is curtly covered in a page and a half. It includes mention of several fires due to "overheating receptacles", without any discussion about the source of heat being the aluminum wire terminations. There is no description of the charred aluminum wired devices that homeowners submitted as exhibits when they testified. The Commission's report does not include any of the many photographs that homeowners and their associations provided of burned out aluminum wired connections, heat-damaged wire insulation, and fire damage. It does not include photos of any of the burned out receptacles that homeowners submitted to the Commission as physical exhibits, nor does the report even state whether they were wired with aluminum or copper. There is little doubt, that they all involved aluminum wire. Field samples (from homes) with damage from overheating copper wire connections are very rare.

On the bottom line, the Commission did find that aluminum wire connections have a substantially higher failure rate than copper wire connections. The report's conclusion for the core issue of reliability and safety reads:

> *… It is unfortunate that, during the decade immediately after it was introduced on a large scale, aluminum should prove less reliable than copper and produce a greater number of failures than copper. The electric system, of course, is so designed that most failures are safe, although they may be disconcerting. If aluminum was less reliable during that period, it would seem logical to suppose that it was also less safe, but statistics are lacking and other evidence is uncertain*

## 15. CANADA

*about the safety aspect. However, the fact that evidence about safety is obscure is understandable: a fire (unlike a failure without a fire) is likely to destroy evidence as to its origin ...*

The Wilson Commission closed up shop without directly facing the question of fire safety. What it calls "failures" are in reality severely overheating aluminum wired connections that destroy the wire insulation and generate a lot of heat. The links between these overheating connections and fire ignition are readily understood from electrical engineering and fire science fundamentals. It is somewhat astounding that the Commission did not share the homeowners' conviction that electrical items heating up, destroying wire insulation, burning up, and arcing in concealed places in the home constitute a safety hazard. The Commission seems to have required some sharper definition, perhaps in foggy scientific terms from one of the industry's experts. As stated in its report:

*The distinction between safety and reliability was not defined clearly in the evidence of any of the witnesses who appeared at hearings held by the Commission.*

Over the course of its investigation, the Commissioner and staff members visited laboratories and organizations in Canada, the U.S., India, the UK, and Japan. In the U.S. they visited UL and Bell Labs, but avoided the CPSC. At the time, the CPSC had completed its initial in-home temperature measurements comparing copper-wired and aluminum-wired receptacles, and fire ignition testing of receptacles from that pilot survey was under way at W-M.

Professor K.D. Srivastava, the Wilson Commission's electrical engineering technical staffer, was in touch with W.H. King at the CPSC, who offered to arrange a visit to W-M for him to observe and discuss the fire ignition tests. King reported that Srivastava declined the offer, explaining that the Wilson Commission was treating the aluminum wiring connection failures as a reliability problem and not as a fire safety problem. The 1975 CPSC document *Hazard Analysis—Aluminum Wiring* is not included in the extensive bibliography of the Wilson Commission report. There is no discussion whatsoever in the report of the CPSC's activity or findings.

The Commission's report comes across as a snow job that promotes the viewpoint of CSA, Ontario Hydro, and the industry. One can surmise that the absence of photos and/or description of the damage done by overheating aluminum wire connections in homes was not simply an honest oversight. If the Commission had been forthright in documenting the destruction of wiring devices and wire insulation that the homeowners were experiencing, it would have been obligated to recognize the safety implications. The fire hazard posed by the destructive overheating would be obvious to the reader, irrespective of conflicting views as to its underlying causes.

In fact, it seems that the Commission itself agreed on that point. Buried in the *Glossary* section at the end of the report, but not discussed elsewhere, is the following clear definition of the fire hazard threshold:

> **temperature, abnormal**: *in functioning electric-wiring devices, an increase in temperature sufficient to cause charring of combustible materials, short circuits, or harmful deterioration of conductor insulation. An abnormal temperature increase may be considered as a sign of a fire hazard.*

The Commission's report was not well received. The homeowner groups did not appreciate that their testimony, exhibits, and concerns had not been taken seriously. They would not be getting the remediation that they sought, and aluminum wiring would continue to be installed in new homes. A homeowner group in Ottawa followed up two months later by suing CSA and Ontario Hydro for $375 million, to provide $1,500 each to rewire all of the 250,000 aluminum wired homes in the province. The group's

## 15. CANADA

founder, R. Jerabek, called the Commission's report "ambiguous, misleading, incompetent, unprofessional and irresponsible."

Nor was the industry pleased, since the report did not really give aluminum wiring a clean bill of health. Wiring device manufacturers were concerned that the frequent mention of overheating receptacles implied that their products were at fault. The Wilson Commission report also contains a list of 43 recommendations to improve residential electrical safety. Some of them would be costly to the industry if implemented. Some implied that the industry was not doing a good job.

One recommendation that was implemented was that Ontario Hydro should offer free inspections for aluminum-wired homes. The inspections, which are primarily visual, have their limitations. A wire connection that looks like new today may be on the verge of overheating tomorrow, or next month, or next year. Electrical and/or thermal measurements are necessary to assess the actual condition of a connection.

In the States, the CPSC measured temperatures at aluminum- and copper-wired receptacles in homes and developed the data needed to answer the question of their relative safety. Although that was exactly the same question the Wilson Commission was charged with investigating, it never took steps to answer it with actual data from Canadian homes.

Later, in 1981, Scarborough Controller Trimmer and Heighington's homeowner group arranged for measurements to be made in three aluminum-wired homes. The abstract of the resulting report states:

> ... *Seventeen (63%) of the receptacles measured reached temperatures in excess of the insulation rating of the attached conductors, with many going substantially higher. The high temperatures reached indicate that an abnormal fire hazard exists in these homes and in other homes similarly wired. ... Two of the homes had recently been inspected by Ontario Hydro. The Ontario Hydro inspection is seen to be inadequate ...*

# 16.
# AN INDUSTRY ON TRIAL

Two aluminum wire trials evolved from the Beverly Hills Supper Club fire. The first one started in December 1979. Plaintiffs claimed that the fire was ignited by an overheating aluminum wire termination on a receptacle in the wall at the telephone receptionist's tiny workspace, her cubbyhole. Defendants countered with an argument that the fire started at a copper wire termination on a different receptacle several feet away on the same wall.

In addition to citing an overheating aluminum wire termination as the most probable cause of the Supper Club fire, plaintiffs' attorneys contended that the defending industry members had worked together to continue the production and marketing of aluminum wire and wiring devices that were known to be hazardous when used together as intended, and that they had failed to warn the public of the hazard.

The list of defendants included major names in the electrical industry. Among the aluminum wire and cable manufacturers were Kaiser, Alcoa, General Cable, Cerro, Southwire, and Anaconda. Among the wiring device manufacturers were GE, Leviton, Eagle, and Slater. Most of the defendants were capable of mounting and sustaining strong legal defense against the allegations. Each defendant was represented in court by a private law firm from the region, backed up in many instances by corporate legal staff.

The first trial took place in the Covington, Kentucky courthouse, only a few miles away from the site of the disaster. Judge Carl B. Rubin presided.

He had already been heavily involved in a mountain of litigation resulting from the fire. Regarding the aluminum wire claims, there were initially about 200 plaintiffs represented by dozens of lawyers. Judge Rubin consolidated the individual lawsuits into a single class action and appointed a Plaintiffs' Lead Council Committee (PLCC) of 14 attorneys to act on behalf of the plaintiffs.

The plaintiffs' attorneys sought maximum compensation on behalf of the victims and their families, including punitive damages. These attorneys worked on a contingency basis, meaning that payment for their expenses and fees would be made from awards and settlements that they managed to achieve. If there were no damage awards or settlements, then plaintiffs' attorneys would not get paid. This is a common arrangement in civil liability cases in the United States. It serves to assure that legal representation for plaintiffs is available even if they do not have the funds to finance the litigation.

Judge Rubin appointed Stanley Chesley as the lead attorney for the PLCC, based on his previous experience. The law firm of Waite, Schneider, Bayless and Chesley was situated in the well-known Cincinnati clock tower building, across the Ohio River from the courthouse in Covington. Their offices became plaintiffs' command post for the first aluminum wire trial. By the time the trial was over, the firm had taken over an additional floor of the building for Beverly Hills exhibits and documents.

When the suit was first filed, plaintiffs' attorneys had little to go on except that aluminum wire had been installed in the club and it had been under attack across the nation as a fire hazard. There was general agreement that the Beverly Hills Supper Club fire was electrical in origin. None of the investigations had yielded a different conclusion.

Total recovery and examination of the physical evidence at the area of origin of the fire was not possible due to the massive destruction and the subsequent moving of debris to locate victims. Most of the electrical cable and wiring devices near the fire's area of origin were not recovered. Sections of aluminum wiring had disappeared, having melted in the heat of the fire.

## 16. AN INDUSTRY ON TRIAL

The specific manufacturers of the aluminum wire and the particular wiring device that plaintiffs claimed to have caused the fire could not be identified.

Normally in product liability litigation, a single product manufacturer is named as defendant. In this instance the entire industry was brought to court, including UL and the manufacturers of the aluminum wire and receptacles. This was in keeping with the plaintiffs' claim that they all acted together to keep the hazardous product on the market. There were 26 defendants at the start of the first trial.

Pre-trial motions were argued. Plaintiffs' right to bring action against the entire industry was upheld. Defendants obtained a ruling that the jury would not be told the names of the defending companies. They also won the Judge's decision to "trifurcate" the trial, meaning that three major issues would be tried in sequence. The jury would first have to agree that an aluminum wire connection failure caused the fire for the case to proceed to the second phase, determining whether UL and the manufacturers did anything wrong. If plaintiffs won on the first two issues, the trial would proceed to the third and final phase regarding liability and damages.

Plaintiffs and defendants wrangled over the documents—what documents the companies were required to give to plaintiffs and how, if at all, they would be presented to the jury. The industry's internal documents were crucial to proving that the companies and UL had acted together improperly.

Outside of the courtroom, Chesley called their actions *conspiracy* and *cover-up*, using words still fresh in mind from the Nixon administration's Watergate scandal. But these terms had narrow definitions for a product liability lawsuit. Chesley sought to get around that problem by calling the industry collusion a *concert of action*. By using this phrase and developing his legal liability theory around it, he hoped to have more flexibility in the presentation of plaintiffs' case. He would not be limited by definitions and legal precedent associated with *conspiracy* and *cover-up*.

Plaintiffs' attorneys worked for two years prior to the trial to gather the information required to present their case to the jury. Prior to the trial, each side had opportunities to probe for the information that it thought was needed to prove its case. This was the "discovery" phase of the litigation. Investigators for both sides examined the ruins and studied the extensive reports, transcripts, eyewitness accounts and investigative reports already available. Depositions were taken of witnesses and some of the defending companies' key employees. The industry documents that the PLCC obtained were thoroughly reviewed and organized.

Finding the key documents and organizing them was a Herculean task. Chesley and his team sifted through mounds of documents, looking for the needles in the haystacks that could prove his claims of known hazard and concert of action. Kaiser, for instance, allowed plaintiffs access to about 60 storage boxes held at its corporate headquarters building in Oakland, California. These boxes contained the company's response to government demands in the CPSC's aluminum wiring case. Chesley, along with several other attorneys and an engineer, spent a day at the Kaiser headquarters reading documents, trying to identify those that might be valuable as evidence. It became clear that a substantial number of "smoking guns" were in this collection, and sorting them out would take many weeks. Chesley had the entire collection copied so that a thorough job of screening the Kaiser documents could be done back in Cincinnati.

Hundreds of documents from the various defendants were eventually tagged as exhibits for the trial. They were in a long row of loose-leaf binders situated near the lawyers' tables toward the front of the courtroom. When court was not in session, anyone could thumb through and read them. People from the CPSC and ABC News were among those who availed themselves of this opportunity.

The documents were potent, but the jury was shielded from the story that they told. Judge Rubin severely limited use of the documents in the causation portion of the trial. He rigorously enforced the rulings that he had made on behalf of defendants; that the trial would first focus only on

# 16. AN INDUSTRY ON TRIAL

the question of causation, and the jury would not be told who the defending companies were. Evidence as to concert of action could not go before the jury unless it first ruled that an overheating aluminum wire terminal ignited the Beverly Hills Country Club fire.

The documents that were offered for use at trial were read and redacted by Judge Rubin and/or his staff. Many documents seen by the jury looked like censored letters from WWII. On any copy that was submitted to the jury, content that identified the defendant companies or touched on the concert of action issue was blacked out.

The first trial lasted 11 weeks. There was a parade of witnesses to the fire, cause and origin investigators and experts on electrical contact technology. This was a high-profile trial, and the courtroom was always filled with attorneys, spectators, and reporters. The courtroom scene often included a full-size replica of the wall at the fire's "area of origin", with the telephone receptionist's cubbyhole, and associated structural features. This exhibit, built by the defendants, was used during trial by experts on both sides to explain to the jury their conflicting theories as to exactly where the fire ignited and how it spread.

There were many interruptions in the trial. Objections and motions took time to argue and decide. Court was in session with the jury present on 22 days during 11 weeks—on average only two days a week over the course of the trial.

The 26 defendants initially presented a solid front. Kaiser and UL were the major players. Their attorneys assumed the leading role in developing and managing the defense. They assured the other defendants that the case was winnable and urged them to stick with it to a jury verdict and not negotiate individual settlement agreements. There were some relatively small companies among the defendants who might be tempted to settle for less than it would cost to continue to the end, and possibly lose.

Kaiser and UL together then abruptly settled, turning their backs on the other defendants. Their confidence in an eventual jury verdict for the defendants had apparently eroded as the plaintiffs' case unfolded in the courtroom. Their expert jury watchers in the visitors' gallery may have sensed that the panel was sympathetic to plaintiffs' case. Other companies settled, following the new lead of Kaiser and UL. Fourteen defendants remained when closing arguments concluded and the question of causation went to the jury.

The jury took less than two hours to return its verdict. The decision was in favor of the defendants—the jury decided that the fire did not start as a result of overheating at an aluminum-wired receptacle.

The second Beverly Hills Supper Club aluminum wire trial came about primarily as a result of what presiding Judge Henry R. Wilhoit, later called "one of the most bizarre occurrences in anyone's memory". A couple of weeks after the first trial had concluded, a truck pulled up in front of the *Cincinnati Enquirer* building. The driver walked into the lobby, handed the receptionist a plain envelope, then left the building and drove off. The receptionist noted the company name on the truck.

The envelope contained a letter to the editor with a title along the lines of *How I Came to My Decision in the Beverly Hills Supper Club Trial*. Its author claimed to have been on the jury and chose to remain anonymous. The letter was quickly brought to attorney Chesley's attention and he lost no time bringing it to Judge Rubin. The letter writer was easily identified from the company name on the truck. He had in fact been one of the jurors.

According to the letter, this juror's house had been wired with aluminum in 1969. While the trial was ongoing, he checked a few receptacles in his house and found that the screws were not loose and that there was no sign of any problem. He concluded that the plaintiffs' experts who testified about aluminum wired terminals loosening and overheating were wrong.

Judge Rubin interviewed the letter-writing juror and determined that he had discussed his personal aluminum wire inspection, and his conclusions, with his fellow jurors during the trial and during their final deliberations. This was confirmed by other jurors.

## 16. AN INDUSTRY ON TRIAL

Jurors in a trial are carefully instructed to base the verdict on the evidence that is presented at the trial. They cannot do their own research or investigations. Plaintiffs' quickly appealed the verdict on the basis of juror misconduct and other factors. Judge Rubin rejected the appeal. He essentially ruled that the law guarantees a fair trial, but it does not have to be a perfect trial. In his opinion the trial had been fair enough.

Plaintiffs' went higher, to the 6th Circuit Court of Appeals, and won. The decision was based solely on juror misconduct, without regard to other substantive issues detailed in the appeal. The jury verdict of the first trial was set aside and a new trial was ordered.

The second Beverly Hills Supper Club aluminum wiring trial started eight years after the fire, on April 30, 1985, in Ashland, Kentucky. The change of location to a distant city, a 2-1/2 hour drive from Covington, was largely motivated by the need to find unbiased jurors who knew little about the first trial and did not have links to the plaintiffs.

Judge Henry R. Wilhoit presided. He joined his family's multi-generational law firm in 1960, and was appointed to the bench 21 years later. Prior to this trial, his court was in an antiquated (1906) Federal post office and courthouse building in Catlettsburg, an old pioneer village. He personally designed the modern court facilities on the third floor of the Carl D. Perkins Federal Building in nearby Ashland. The second aluminum wiring trial started the day after that brand-new building opened to the public.

Space was made available for the opposing parties to use for the duration of the trial. Chesley and his team occupied a large room on the second floor of the courthouse. Documents and exhibits were relocated from Cincinnati to this "war room" for ready access during the trial.

Plaintiffs also rented a furnished apartment across the street from the courthouse, above a bar and grill, as a sort of off-site office and "break room".

One can only imagine what some of the neighbors thought of the stream of people flowing in and out of the apartment's stairwell entrance door next to the bar and grill. There were some regulars, some one-timers, many men and a few women.

The 14 defendants vowed to hold the line. Their confidence had been buoyed by the favorable decision of the first trial. There would be no premature settlements this time. The case was winnable.

With one major exception, the second trial proceeded along same path as the first. In the second trial, the questions of causation and concert of action were tried simultaneously. Testimony and documents were presented that had not been allowed to come before the jury in the first trial. But when that phase of the trial ended, the jury would deliberate the issue of concert of action only if it first found that an aluminum wire connection failure caused the fire. Beyond that, there was essentially the same parade of witnesses, exhibits, documents and expert testimony.

Well into the second trial, the defendants asked the court to dismiss the case, claiming that there was no aluminum wire in the vicinity of the receptionist's cubbyhole where plaintiffs placed the fire's point of origin. Until that time, including the first trial, no direct evidence had been submitted as proof.

Chesley had a limited time to respond. In the war room, his team frantically searched through thousands of photos from all sources looking for any that clearly showed aluminum wire remains located where the cubbyhole had been. One photo was finally found, and the trial resumed.

Once again, about halfway through the trial, the "all for one and one for all" defense collapsed. Leviton agreed to settle. All parties met in Judge Wilhoit's chamber and Leviton's settlement was approved. The remaining defendants argued that it would prejudice their case if the jury found out that there had been settlements. They reasoned that the jury might surmise that there had been settlements if they saw a diminishing group of defense attorneys in court. Judge Wilhoit agreed. He ordered that the settlement was not to be publicly disclosed and that lawyers representing companies

## 16. AN INDUSTRY ON TRIAL

that settled would be present in court for the balance of the trial. The conference ended.

Less than an hour later, a newspaper reporter approached Chesley in the hallway and asked if it was true that Leviton had settled. Chesley notified the judge that his secrecy order had apparently been breached. Judge Wilhoit summoned the plaintiffs' and defendants' attorneys back to his chamber. He instructed them to attempt right then and there to develop a settlement agreement for all defendants that would bring the trial to an end. The defense attorneys assured the Judge that they had the authority to negotiate and the process moved very quickly to an agreement.

The last step toward achieving this universal settlement was to get legal and executive approval from the defending attorneys' corporate clients. These were the days before cellphones, and there were only a few telephones in the building available for the attorneys' use. The courthouse halls came alive. Attorneys searched for desk phones that could get an outside line and some waited to use pay phones in the hallways. Witnesses and staffers in the building tried to be discrete and keep out of earshot. Attorneys using the public pay phones looked over their shoulders to make sure nobody was eavesdropping.

All but one of the defending companies approved the proposed settlement package. GE stuck to its corporate policy of never giving in on an aluminum wiring issue and refused to settle. To his chagrin and embarrassment, GE's principal defense attorney in the Supper Club litigation had been overruled by the company's management and in-house lawyers. The attempt to achieve a universal settlement failed. The trial resumed.

More defendants settled as the trial progressed, and Chesley actively sought to negotiate with those that remained. By the time closing arguments by plaintiffs and defendants concluded, the only defendant left was GE. The company had refused all settlement proposals. Settlements by the 13 other defendants in the second trial totaled a bit more than $4 million.

On July 15, 1985, the jury deliberated its verdict on causation. It was reached in less than an hour. The jury agreed with plaintiffs, deciding that an overheating aluminum-wired receptacle terminal ignited the Beverly Hills Supper Club fire. GE had bet heavily that the first trial's causation verdict would be repeated. Now that it lost that bet, the company was in a very poor negotiating position.

Next, the jury would retire to consider the *concert of action* question. Judge Wilhoit was about to provide the panel with an explanation of the applicable law when GE agreed to settle for $10 million. The company's intransigent policy—not to give an inch on aluminum wire issues—had cost it dearly.

It is tempting to feel sympathetic toward GE if one considers it to be just a poor receptacle manufacturer that got dragged into the aluminum wiring issues by the actions of others—Kaiser and UL in particular. That image is not quite correct. GE had been a wire and cable manufacturer. In 1947, it was among the first to introduce aluminum building wire into the market. The promotional flyer that launched its aluminum building wire product line noted the post-war copper shortage and claimed that aluminum was a satisfactory substitute. The flyer bragged that GE's aluminum building wire products were UL approved, but there is no mention of the fact that the UL listing was granted only on a temporary emergency basis because of fire safety concerns. GE had long since given up producing aluminum building wire at the time the company settled in this trial. But General Electric Supply Company, its subsidiary, marketed Kaiser's KA-FLEX aluminum NM cable for house wiring.

When the trial was over, Judge Wilhoit conducted an informal session for the jurors to ask questions. They were impressed by the ease with which plaintiffs' lawyers accessed the documents and exhibits. They asked to see plaintiffs' war room, and the request was granted. Judge Wilhoit later described the room as follows:

## 16. AN INDUSTRY ON TRIAL

*The PLCC operated out of a 40' x 40' room on the second floor of the courthouse. This room ordinarily would be quite sufficient, but it was literally stacked with boxes of documents, exhibits, desks, and work tables complete with a grimy-looking hot plate and coffee pot. This was the scene for all of the after-Court hours that were always required.*

He also noted that:

*From literally thousands of documents, the PLCC demonstrated considerable skill with their selection and presentation of several hundred of them that were eventually published to the jury. Without further burdening this opinion, suffice it to say that the document presentation got the plaintiff class to the jury. ... It was a "documents" trial for the most part.*

And, speaking of the total settlements ($43 million) that the PLCC had gained on plaintiffs' behalf from the aluminum wire, PVC, and other class action lawsuits, he wrote:

*Although substantial sums of money have and will be distributed to individuals and families touched by this terrible tragedy, it could never be sufficient to replace the losses that have occurred. It is only the best that we can possibly do.*

*The Attorneys that have represented these parties are to be highly commended for their tireless efforts.*

# 17. ARRESTED DEVELOPMENT

> *But, as with so many persistent puzzles, the resolution does not lie in more research within an established framework but rather in identifying the framework itself as a flawed view of life.*
> —(Stephen J. Gould)

WRITING FOR *NATURAL HISTORY* MAGAZINE, GOULD MIGHT JUST AS WELL have been discussing the persistent aluminum wire connection problem and its associated fire hazard. Virtually all of the UL and industry effort over the past half century has been expended within the confines of a narrow framework of "heat cycle" testing. The flaw in the established framework is that heat cycle testing does not effectively determine whether a connection type will last and perform safely in homes.

This might best be understood by analogy. Suppose that medical providers agreed that a treadmill test is the only examination needed to determine a patient's long-term health prospects. Hypothetically, the decision was reached after hundreds of meetings and much debate and compromise on the details; speed and incline of the belt, timing and number of on-off cycles, examination room temperature, humidity, air circulation, measurements to be made—pulse, blood oxygen level, body temperature, heart rate, and such—and what constitutes passing the test. But no matter how carefully the test is defined and standardized, it would be blind to many lethal health threats.

Similarly, the tests that UL and the manufacturers settled on in the early 1970s for aluminum building wire and its connectors are also limited in scope. At that time, a quarter-century after UL President Alva Small warned about the potential fire hazard of aluminum wire in buildings, and as the number of homes so wired exceeded two million, UL initiated its first tests intended to sort wiring devices that would work well with aluminum from those that would not. UL was under considerable public pressure to either solve the problem or withdraw its approval of aluminum wire for branch circuits in homes.

The foundation for the new test program was the existing UL standard for connectors, UL486, which was already known to be inadequate. Aluminum wire connections made with connectors that UL listed under that standard were failing in service. That is what compelled Kaiser to adopt its own tougher test requirements in the mid-1950s, and led to the UL-NEMA field failure survey of the mid-1960s.

The existing standard used a heat cycle test to gauge the connection performance. Higher-than rated current was applied in an on-off cycle while measuring connector temperature. UL and the manufacturers agreed to base the new test program on a heat cycle test employing higher current and many more on-off cycles.

This new "kick the tires" test is seemingly severe. But, as with the treadmill test described earlier, it is not thorough. No matter if the tires are kicked more often and harder—more cycles at higher current—a heat cycle test does not reveal certain design and material weaknesses that result in aluminum wire connections deteriorating and overheating in homes.

UL and the manufacturers expended considerable effort going down a narrow path of inquiry and experimentation that eventually lead to agreement on the details of current, on and off time, connection assembly methods, number of specimens, pass/fail criteria, and more. No effort at all was made to determine if favorable test results would assure safe long-term operation in homes. Manufacturers set up their own tests to see whether

## 17. ARRESTED DEVELOPMENT

their wire or wiring device would pass the new requirements. If they didn't, either the product design or the proposed test requirement could be changed. As eventually agreed to, the applied conditions of the new test were in some ways more aggressive than those of the older standard, but its pass/fail performance requirements were weaker.

Branch circuit wire connections are intended to be permanent. There are hundreds of them in every house. Each and every one of them is expected to perform safely for an indefinite time without periodic maintenance or replacement. They should last as long as the wire does. Connections in copper-wired homes generally meet that expectation. Many common connection designs in aluminum-wired homes do not.

There is also an important difference in severity of failures. Overheating aluminum connections pose a higher risk of fire ignition than overheating copper connections under the same conditions. Failing aluminum connections continue to be functional in the circuit for a longer time, generating heat whenever current flows. Whatever is plugged into the circuit that draws the current operates normally. There is no overcurrent that would trip a circuit breaker or blow a fuse. Undetected, overheating connections in the circuit destroy the adjacent wire insulation, sometimes causing short circuits. Occasionally, overheating aluminum wire connections glow red-hot.

Fire safety considerations dictate that the test procedures used to qualify components of the permanent built-in wiring system in homes must approve only the combinations that will provide safe long-term performance and reject all others. In 1970, when the new heat-cycle test was being debated and refined, the fundamentals of designing test programs to accomplish that sort of objective were already known. A combination of stress tests and life tests would be required, along with strict pass/fail limits and statistically sound sample size for the various tests.

The test program that UL and industry committees settled on does not include life testing. The stress test that it relies on, the heat cycle test, does not relate to actual service conditions for an aluminum wire connection in a home. Connections can deteriorate substantially under the applied conditions of the test yet still be accepted for UL listing.

The effectiveness of the test was challenged even as its procedures and pass/fail limits were being firmed up. UL's VP W.A. Farquhar deflected suggestions that the effectiveness of the test needed to be validated. In a letter to H. Cook of the Aluminum Association, recounting a meeting at NEMA headquarters, he wrote:

> *During the discussion a question was asked whether we would then run a study to determine the adequacy of the test program. I suggested that rather than do this we indicate our willingness to accept submittals of products intended to be qualified for use with aluminum and that we make it clear to the submitters that we were keeping the program open ended with the idea that if any strange or unusual results occurred we could then explore in a different direction to make sure that we really had a product which was suitable.*

Based on Farquhar's previous actions, "strange and unusual results" should be taken to mean failures in homes. The proof of the new qualification testing was to be done by homeowners. If failures occurred in homes with wire and devices that UL listed using the new test, then Farquhar would "explore a different direction".

Strange things did actually occur in the lab tests. Inconsistent results were obtained in UL's own testing. There were unpredicted irregularities.

One such instance occurred in the testing of receptacles for aluminum wire with the new CO/ALR rating. Designs submitted by several manufacturers were being tested at UL's Melville, NY facility. A significant jump in connection operating temperature occurred across all samples immediately after a shutdown of the building's electrical system for some planned maintenance. The responsible engineer at Pass & Seymour, whose receptacles were among those being tested, reported that:

## 17. ARRESTED DEVELOPMENT

> *... at 418 cycles our device began to run above the allowable 100C temperature rise. At the same time, all devices of other manufacturer's submittals on test with ours began to show a significant rise in temperature. This presents some question as to the validity of UL test procedure ... This latest development, plus other information received in the last few months, raises some question as to the reliability of any wiring device of present configuration for use with aluminum wire.*

The surprise was that ten days of sitting idle in a lab room at zero current was a more aggressive stress test for those aluminum wire connections than ten days of cycling at high current. A similar incident occurred at Kaiser's research lab. A cluster of 14 failures occurred shortly after a month-long shutdown of the heat cycle test system. Five of the new failures occurred in the first current-on cycle when the system was restarted.

The culprit in these curious incidents was just the normal moisture in the air. The high temperature of connections undergoing the UL heat cycle test suppresses normal moisture-related deterioration processes. Researchers at Battelle were among those unsuccessfully calling for a more realistic and effective test procedure framework, writing:

> *Closely related to this work is the development of accelerated but <u>realistic</u> testing techniques. Methods must be developed which in a relatively short period of time will subject connections to a realistic thermal, mechanical, and environmental equivalent of the expected service life.*

On the other side of the ledger, with current and temperature way out of the ballpark relative to normal use in homes, the heat cycle test induces failure processes that would not normally be active in actual use. In late 1975, Kaiser's G.H. Kissin wrote:

> *The next point we should consider is whether cycling tests under overload conditions can be realistically related to actual service performance. There is a possibility that overload cycling brings into play mechanisms that are not at work at normal rated current loads. Thus, we conclude that a substantial portion of our work should conform with Professor Gatos' recommendation that we do our cycling at rated current loads and introduce variables which might cause disturbance of normal functioning ...*

Another surprise occurred during UL's listing tests for Leviton's new CO/ALR receptacle. One of the six samples unexpectedly failed. The product was approved nevertheless. The failure was disregarded, on the basis that the failed wire termination might not have been properly tightened.

Also, during UL testing, CO/ALR receptacle designs that passed the test with one brand of aluminum wire failed when tested with another. Discussing the particular wire to be used for wiring device testing, UL engineer E.W. Krawiec wrote:

> *... Both the "old" Kaiser material and the ECH-24 material induced failures in CO/ALR devices. However, the type of failure induced by the Kaiser material was of a more severe nature ... thermal runaway/burning as opposed to exceeding an arbitrary maximum temperature rise.*

These incidents support the contentions of those who, at the time, challenged the effectiveness of UL's new test procedures. On the other hand, Farquhar's reported objective—to improve performance—was met to some extent.

The new standard test for wiring devices fell short of providing a complete solution to the aluminum wire connection problem. The causes of

## 17. ARRESTED DEVELOPMENT

failure were not well understood. Aluminum-wired CO/ALR receptacles and switches have subsequently failed in lab tests and in homes.

At about the same time, in the early 1970s, UL conducted another program to weed out the worst-performing varieties of aluminum wire. Its Aluminum Alloy Assessment Program followed the same narrow framework of heat cycle testing. A hodgepodge of variables was experimentally explored. Aluminum wire samples were tested with different wiring devices, different terminal screw materials, cycling at various current levels, and different terminal screw tightness. Twenty-four aluminum wire variations from several manufacturers were evaluated, differing in mechanical properties, composition and processing history. Copper wire was tested alongside the aluminum wires. None of the aluminum alloy wires equaled the failure-free performance of copper wire.

Test results for the aluminum wires were inconsistent. Trends that were anticipated did not materialize. T.J.D'Agostino, the project engineer, wrote:

*The primary objective of the aluminum alloy assessment program was to relate connectability (termination performance levels) to the mechanical properties of the conductor material. ...*

*... The results ... were not realized for most all of the alloys ...*

*At the April 11, 1972 meeting of participants ... the data was reviewed concerning degrees of correlation. ... approaches to defining performing alloys were also discussed. However, each attempt to relate physical properties data ... proved fruitless as each approach presented exceptions in the data to what appeared to be significant factors. ...*

An arbitrary allowable number of failures was eventually chosen as the pass/fail limit, under which several manufacturers' alloy wires were accepted for listing by UL. By late 1973 they were in the marketplace and, with one exception, the poorly-performing EC grade aluminum NM cables were no longer being manufactured.

The exception was Kaiser KA-FLEX, an EC grade aluminum wire that failed to make the cut under the original pass/fail limits. Under pressure from Kaiser, Farquhar agreed to change the pass/fail criteria so that it would pass. The manufacturer then continued to produce the same wire under a new brand name, KA-FLEX ALR.

The claim is often made that aluminum wiring installed during or after 1972 is safe because of the changes in UL's approval procedures. But the performance of the wire has no relationship to the home's date of construction. Even after the newly-approved alloys appeared in the market, the old material continued to be installed in new construction until the existing inventory was all consumed. Kaiser's poorly-performing wire, new brand name and advertising hype notwithstanding, continued to be produced until the aluminum NM Cable market totally collapsed in the late 1970s. Some of it still awaited purchase from distributor shelves in the mid-1980s. The so-called "new technology" CO/ALR receptacles were not widely distributed, so ordinary receptacles were used, with UL's nod of approval. On the positive side, Southwire's EEE alloy, which performed almost as well as copper in the UL tests, was already on the market in the late 1960s.

There is no such thing as simply "aluminum wiring". There are aluminum wires with relatively thick and tough insulating oxide on the surface, and aluminum wires with a somewhat conductive surface. There are soft aluminum wires and hard aluminum wires, there are EC aluminum wires and alloy aluminum wires, there are some that are more creep resistant than others. And those are only variables associated with the wire. A menagerie of additional variables is associated with the wiring devices and connectors. On average, the new aluminum wire alloys demonstrated improved performance in the UL tests, but the technical reasons were not well understood.

The production process variables were known to influence connection performance. The hoped-for relationships between physical properties, alloy

## 17. ARRESTED DEVELOPMENT

composition, and test performance did not materialize. Without that, there was no quick, easy and economical way for UL inspectors to periodically check the product coming out of the factories. UL proposed that the manufacturing process had to be specified and monitored. Kaiser researcher C.R. St. John wrote:

> ... *The UL people were dismayed to find that the best wires had different composition and different fabrication histories. Wires of similar composition showed different results. Due to this, then, approval is contingent upon the wire always being made by the process used to produce the samples.*

The UL alloy and wiring device projects resulted in the promise of some improvement for new installations. Attention turned to producing and marketing the new products—wiring devices marked CO/ALR and branch circuit aluminum NM cable with a new UL label. After the new test procedures were adopted, several attempts were made to develop and validate a theory linking the results to actual service life in a house. They failed. Manufacturers funded a major effort at Battelle to develop and validate a theory that would connect the test results obtained at different current levels. That also failed.

In aluminum-wired homes, in the shadow of the spotlight on receptacle terminals, wire and cable connections of all types were also overheating and burning out. In any home, there are numerous twist-on connector splices and wire connections to circuit breakers. Overheating failures of these connections also are common when aluminum wire is involved but rare when they are copper-wired. The need for improvement was recognized in the 1960s, but more than a decade passed before UL acted to improve its applicable standard and test procedures.

UL and industry committees eventually agreed on a more rigorous test procedure for wire and equipment connectors. Again, some improved performance was achieved, within the limited framework of a heat cycle test. But, still, kicking the tires more often and harder—more cycles at higher current—fell short of what was needed to assure safe long-term life in a home.

In about 1995, the Ideal #65 "Twister", listed by UL under its new standard, entered the market. Essentially, it is similar to one of the company's existing twist-on splicing connectors but it is filled with a corrosion inhibitor grease. There was a ready market for it. At the time, no twist-on connector was UL listed for the aluminum wire pigtailing repair, and the CPSC's recommended connector was expensive and not available in many parts of the nation. Within two years of its introduction, overheating failures of the Twister connector were reported in homes and laboratory tests. The CPSC tried, without success, to convince the manufacturer to discontinue its promotion of the Twister for whole-house aluminum wire pigtailing repair.

Shortly after the Twister connector entered the market, more than 4,000 of them were used for aluminum wire pigtailing at a luxury condo complex in Corpus Christie, Texas. Eighteen burnouts of these connections occurred within the first year after installation. At the site for a meeting on the problem, Ideal's engineer replaced one of their Twister connectors that was running hot. His newly-made connection then also ran hot. The manufacturer and UL blamed the failures at the complex on "poor workmanship". After another year and more failures, the condo management removed all of the Twisters and replaced them with the COPALUM crimp connector.

The COPALUM connector is recommended by the CPSC in its Publication #516, "Repairing Aluminum Wiring". It has been used for that purpose for more than 40 years, with no failures reported. A second connector, brand name ALUMICON, was later recommended by the CPSC as a backup if the COPALUM was not available. It has now been used for pigtailing repairs for more than 10 years without any failures reported. These

## 17. ARRESTED DEVELOPMENT

connectors were recommended by the CPSC after passing tests somewhat different than the UL tests and with stricter acceptance criteria.

UL's wiring device and connector test procedures of the 1970s were eventually formalized and published as new UL standards. But UL did not follow through and standardize its connectability test for the aluminum wire. Instead, UL now simply accepts any aluminum wire that is registered as an AA-8000 series alloy with the Aluminum Association. There is no longer a minimum connection performance test for the wire, and no need to control the production process. The manufacturers just apply to the Aluminum Association for a registration number for each proprietary alloy that they develop and UL accepts it without any connectability testing.

The registration of an aluminum wire alloy with the Aluminum Association is like the registration of an automobile. Being registered does not assure that a vehicle is safe for use on the highway. The AA-8000 series registration does not mean that the wire has been tested and certified to be safe to use for wiring a house. Today, neither UL, the Aluminum Association, nor the manufacturer has any obligation to test connectability of the registered alloys. If he were still alive today, Leviton's chief engineer M.J. Weitzman would no doubt label the AA-8000 designation a "marketing gimmick", as he did regarding the AL-CU mark on receptacles.

There is no question but that Southwire's EEE alloy performed well. Southwire deserves credit for introducing it in the late 1960s and providing test data supporting its claim of superior connectability. Southwire's results were also confirmed by others in subsequent tests. It was a step in the right direction, but still fell short of equaling copper wire performance. Other manufacturers developed their own alloys, attempting to replicate the physical properties and performance of EEE without impinging on Southwire's patents. Seemingly unknown to all at the time, and outside their narrow framework of experimental exploration, is that some samples of Southwire's EEE alloy wire have an electrically conductive surface, not the assumed insulating oxide.

The electrically insulating oxide on the aluminum wire surface was the "elephant in the room", known to be a major factor but largely ignored. In the 1970s, many of those involved accepted the fact that the aluminum wire connection problems could not be reliably overcome without addressing the surface oxide properties. Kaiser researcher C.R. St. John, for example, wrote:

> *We have concluded in Aluminum Research that the most probable success, in solving building wire connection failures, will come from a process which lowers the surface contact resistance. ... An alloy with increased resistance to creep is a step in the right direction, but an alloy will not solve the oxide problem.*

W.H. Abbott at Battelle also recognized the importance of the oxide properties, writing:

> *Battelle believes that the nature of the surface film on the wire is the dominant factor leading to failure.*

And even UL's VP W.A. Farquhar wrote:

> *There are entirely too many field failures where aluminum has been employed as the conductor. ... There has been considerable talk of new alloys which are much more resistant to creep. There still, of course, is no good answer for the oxide problem short of coating the conductor with copper.*

Farquhar was referring to the copper-clad aluminum wire that Texas Instruments had developed and was promoting. He clearly considered it to be a solution to the aluminum wire connection overheating problem, but he did not promote its use. Instead, Farquhar clung to the hope that the

## 17. ARRESTED DEVELOPMENT

wire, wiring device and connector manufacturers could find a way around the oxide problem.

UL did not try to determine the relationship between the oxide properties and the test results. Other than measuring the current in the heat-cycle testing, no electrical measurements were regularly made. That is somewhat astounding, considering that an electrical contact interface is at the heart of the problem. The health and performance of the test specimens could have been monitored and understood through basic electrical measurements and technical knowhow of the time, but they were not. This is analogous to monitoring the body temperature but not the pulse rate of a person on a treadmill test.

Electrical fire safety took an additional step backwards, in that same era, when manufacturers of a broad range of electrical items climbed on board the aluminum cost reduction bandwagon. Westinghouse, for instance, initiated a program called "ABC"—*Aluminum Before Copper*—for employees making decisions on the metal used for conductive components of its electrical products. As with aluminum wire, long-standing design and testing procedures that had worked for copper components had only mixed success with aluminum. It was left for the end users to sort out. Some of the substitutions were successful, others were not. When they were not, the problem was most often at the contacts and connections involving the aluminum components.

Aluminum appeared as a cost-cutting replacement for brass and copper in products ranging from light bulbs and sockets to locomotives. Many substitutions were troublesome. Lighting problems led Chrysler to caution against using aluminum-base bulbs. Both ALCO and GM ran into costly aluminum connection failure problems in railroad locomotives that they manufactured.

For home owners and occupants, failures and fires have occurred at electrical service panels and meter bases. There are multiple current-carrying connections involving aluminum in many of today's circuit breaker panels, some of which are failure prone. A report issued by the Canadian Electric Association at the end of 1975 notes that a majority of residential electrical fires during the period studied started at service entrance and switch panels, and that the incidence of such fires increased in the 1970-75 timeframe. That correlates with the increased usage of aluminum current-carrying elements in the panels that started a few years prior.

The potential benefits of using aluminum as an electrical conductor are still there. Counteracting that is the cost and safety risk of aluminum connection failures. The progress that UL and industry made in the early 1970s to reduce the failure rate represents a step in the right direction but it came to a halt far short of achieving the hoped-for level of safe copper-equivalent connection performance in actual service.

By 1975, the framework of UL's new test procedure was locked in. It has changed little in the half century since then. The need to go higher on the ladder of electrical safety has been amply demonstrated. But UL and the manufacturers paused on that ladder, and have consistently rejected proposals for substantive change to make the test procedures more effective. That leaves us in a sustained state of arrested development, aptly described by NBS researcher J.Rabinow in 1975:

> *Summarizing, I think that the combination of presently available aluminum wire and connectors provide a form of Russian Roulette where, under the best conditions, one may expect no difficulty but where under other conditions one can be sure to have problems.*

"Problems" means that there will be some fires, such as described in an affidavit submitted to the CPSC in 1977 by P.M. McDonald:

> *I am Fire Marshall for the Orange County Rural Fire Prevention District #3, Orange County, Texas. I have been a member of the Fire Department in Orange County for ten (10) years.*

# 17. ARRESTED DEVELOPMENT

*During the course of my duties as Fire Marshal, I investigated the January 23, 1973 mobile home fire involving the property of Mr. and Mrs. William Jardell.*

*My investigation revealed that the apparent source of the fire was a short in an electrical wall socket which had been wired with aluminum wiring. ...*

*On the night of the fire the Jardells awoke to the smell of smoke. Mrs. Jardell went for help while Mr. Jardell rushed to save their children, William Jardell, II, age 7, and Elizabeth Jardell, age 3.*

*As a result of the fire both children received burns from the intense heat inside the trailer. Elizabeth was injured seriously, receiving burns on her face, arms, stomach, legs, and back. Both children have undergone plastic surgery for their burns and are still being treated by the Shriner's Burn Institute in Galveston, Texas.*

*Mr. Jardell died from injuries received in the fire.*

# 18.
# THE MYTH OF THE SELF-REGULATING INDUSTRY

*Historically, the fire death rate in the United States has been higher than most of the industrialized world. This has held true for both fire deaths and dollar-loss rates. The causes of the United States prominent standing in this area are not entirely clear and have been the subject of debate for some time. To compound the issue, the United States is comparatively safety conscious and one of the most technologically-advanced nations in the world. To have such high fire death rates is perplexing for a country that ranks so highly in those two areas.*
—(US Fire Admin.)

FOR FIRES OF ELECTRICAL ORIGIN IN HOMES, IT IS LESS PERPLEXING. THE electrical safety net, presumed to be competently maintained by a self-regulating industry, has some big holes in it.

Residential fires in the U.S. consistently account for more than half of the annual fire fatalities, injuries and property damage. Fires of electrical origin constitute about 10% to 15% of those losses, with around 50,000 fires and 400 to 500 or so fatalities each year.

In stark contrast, U.S. commercial passenger aviation achieved a remarkable safety record of 10 consecutive years without a fatal accident. At the same time, passenger traffic increased and the cost of air travel went down. Back when the UL and industry committees were wrestling with the aluminum wire problems, there were kiosks in the airport terminals selling flight accident insurance to departing passengers. Crashes occurred from time to time, as they did in travel by any mode.

When comparative airline safety data became commonly available, many passengers showed a preference for the safer airlines, even if the fare was higher or the schedule less convenient. Delta benefitted from its top safety ratings, while some low-ranking airlines went out of business. An "all hands on deck" effort by industry and government created a safety-oriented organizational structure and culture that led to today's outstanding airline safety level. We have also seen how, for instance with the Boeing 737-Max, relaxing the tight system of safety-oriented checks and balances can have disastrous consequences for all concerned.

A revolution in automobile safety was gathering steam at the same time that aluminum building wire was taking hold. Highway death statistics were in the news in the late 1950s. Automobile manufacturers were initially unresponsive, convinced that style sold cars, that any mention of safety depressed sales and that safety improvements would cost too much. Coincidentally, Ralph Nader's book, *Unsafe at Any Speed*, came out in 1965, the same year that Kaiser introduced its KA-FLEX aluminum NM cable for branch circuits in homes.

In 1972, while UL was still in the middle of its first substantive aluminum wire connection tests, Lee Iacocca, President of Ford at the time, famously said, "safety doesn't sell". He was proven wrong, and the upward trend of highway carnage was reversed. Today, customers are eager to get—and pay a premium for—cars with high crash test ratings and advanced safety features.

The safety advances in air and automobile travel have been nurtured by government participation in the process of setting safety objectives, developing standards and providing some of the resources required to accomplish the goals. Essentially, the government represents the public interest in that process, and sometimes dominates the process.

## 18. THE MYTH OF THE SELF-REGULATING INDUSTRY

There was no significant government participation in the process of developing the safety standards for components of the aluminum building wire system. The process was administered by UL and was accomplished in cooperation with the product manufacturers. They were considered to be the stakeholders—those having a stake in the outcome. The general public—the eventual customer, end user and party at risk—was not considered to be a stakeholder.

The electrical equipment manufacturing industry has been "self-regulating" since its inception in the mid-1800s. This is generally taken to mean that it is largely free of government oversight and intervention. But absence of government involvement is only one aspect of self-regulation. A self-regulated industry or profession is presumed to have its own codes and standards for organization, procedures, behavior, ethics, auditing, accountability, transparency and/or other appropriate aspects of the particular business.

No such internal regulatory framework is in evidence for the sector of the electrical equipment industry responsible for the aluminum building wire system. There is no code of ethics or behavior that guides member companies or their employees regarding the relative priority of public safety versus commercial interest. There is no product safety mandate. Insofar as public safety is concerned, for this sector of the electrical equipment industry, for this product family, self-regulation is a myth. The sector is unregulated, not self-regulated.

The accounts of the April 15, 1971 meeting at Travelers Hotel (Chapter 5) bear this out. They paint a picture of the wild west of electrical fire safety—a gunfight on the muddy street pitting commercial interests against public safety concerns. Safety concerns were outgunned. UL's VP W.A. Farquhar, who called the meeting and presided, does not appear to have set down any objective other than deflecting aluminum wire bans and Federal Government intervention. He apparently was not bound by any UL or industry self-regulatory requirement to do otherwise.

In part, the UL and industry actions and inactions may be attributed to—but not excused by—a mindset that accepts a certain level of overheating failures and fires as being normal. Reducing electrical fire losses from equipment and wiring failures may require a different approach, one that continuously identifies and eliminates existing systemic failures and proactively avoids new ones by eliminating weak points in applicable standards. Perhaps that is too much to ask of an unregulated industry that is well set in its ways.

The only user representative involved in the UL and industry standards development meetings was Dupont engineer, M.M. Gilbert, who did his own connection testing to explore the possible use of aluminum conductors at the company's facilities. Writing to UL's W.H. Hoffman at the end of 1970, he stated:

> ... *According to your reference letter, the ad hoc committee ... to determine the direction to evaluate wire-connector and wiring device terminals is composed of personnel from companies who have been telling me for the past two years—there is no problem.*
>
> *I guess after the frustrating past two years, it is difficult for me to believe these same people can constructively advance the art of aluminum wire and its terminations.*

"These same people" that he wrote about were the ones who determined what was produced and approved for installation across the country under building codes and laws that rely on adoption of the National Electrical Code (NEC). The practical application of the NEC requires that we have faith in the standards and the lab testing procedures. As NEC engineer J.H. Watt wrote:

> ... *As you may know, Section 110-14 (a) of the 1971 NEC states "Terminals ... used to connect aluminum shall be of a type suitable for the purpose." Accordingly, we must rely on standards, testing*

## 18. THE MYTH OF THE SELF-REGULATING INDUSTRY

*procedures, and listings of recognized testing laboratories to achieve compliance with the objective of this code rule.*

This faith-based approach did not work well for homes with aluminum wiring. The risk of fire was not taken seriously by those involved in the development, approval and marketing of the wiring system's components. Many problems were swept under the rug.

To illustrate, a 1972 UL product listing report for one brand of "new technology" CO/ALR receptacle states that five samples passed the heat cycle test. A supplementary page, marked "not for outside distribution", explains that six samples were tested, but one suffered a "catastrophic failure". The failure was whitewashed away and the manufacturer's CO/ALR receptacle was accepted for listing by UL. Since then, failure of aluminum-wired terminals on these new technology receptacles in homes and laboratory tests have been reported.

*The mistake is that you cannot assume that people act responsibly just because they are in a position of responsibility.*

# POSTSCRIPT

Stan Chesley finally arrived for our meeting—five hours late. The lead plaintiffs' attorney in the Beverly Hills Supper Club class action lawsuits had a stack of documents to review before deposing an opposing expert witness the next morning.

We were both weary from travel. I had driven ten hours in foul weather. He had made a round trip by air to get the documents, encountering flight delays due to the same storm system. The original plan was to meet for supper at 5:00 and review the documents together. These were the first industry internal documents related to aluminum wiring that either of us had ever seen.

Chesley first contacted me by telephone in 1978, a year before that meeting. At the time I was heading up the aluminum wire test work being done at Wright-Malta Corp. (W-M) for the CPSC, and had been engaged as a technical witness on behalf of the Hersh family. Chesley wanted me to serve as an expert witness for plaintiffs in his case. By that time, the tests that we were performing at W-M had clearly demonstrated the aluminum wiring problems and hazards. I told him I could provide testimony on the technical analysis and on any of my test results that were publicly released by the CPSC.

But, in that first phone call, Chesley wanted my participation beyond the technical aspects. He intended to prove that there was "conspiracy" and "coverup" by UL and industry. I did not believe it at the time. In fact, having worked until then in the corporate world, I had a negative view of product liability lawsuits and the lawyers who made a living from them. I previously worked for GE and IBM and knew that there were competent people in the companies who did their best. I told Chesley I did not think I could be of any help in that regard. Yes, there was a product safety problem with aluminum wiring, but in my view—at that time—it most likely resulted from honest oversight rather than intentional unprofessional actions. "You'll see," Chesley responded, "when I get the documents."

A week or so after that first conversation, I was staring at the remains of the Supper Club. Chesley had successfully argued for a court order postponing demolition and cleanup until plaintiffs' experts had ample opportunity to complete their on-site work. I did not know what to expect, was shocked by the devastation and had no idea how I could help in the investigation. I decided to simply explore the ruin and photograph all the evidence I could find as to where and how aluminum building wire had been installed. (Seven years later, at the second trial, one of my on-site photos was a crucial piece of evidence. It provided proof that aluminum wiring was at the area of origin and countered a defendants' motion to dismiss the case.)

When we met to go over that first trove of documents, Chesley said he was very tired but had to be mentally alert and agile the next morning for the deposition. He needed as much sleep as possible. Sometime after 11:00, we agreed that I would read through the pile of documents, select any that he should review before the deposition, and we would meet for breakfast at 7:00 in the morning to go over them. But I was also tired, and did not believe I could actually get through even a quarter of the stack of reports and memos without dozing off.

How wrong I was. I finished reviewing the entire collection at about 4:00 in the morning. It was an eye-opening and energizing experience—the book that you just can't stop reading. My naïve confidence in the self-regulating electrical industry was shaken. I was angry. I knew about some of the tragic fires. How could all this be swept under the rug?

At breakfast, I told Chesley that the documents—the words of the people inside the industry—showed that he was correct. Whether *coverup* and *conspiracy* were exactly the right words could be debated, but they

## POSTSCRIPT

reflect the historical reality to a reasonable extent. How did he know that before ever seeing any of these memos and reports? He explained that his hunch was based on several issues he had previously litigated. Different technologies and different companies, but there is a certain predictability as to the behavior of people within the companies who are involved when embarrassing problems arise.

The stack of documents became a mountain as the case progressed. A solid foundation emerged for Chesley's theory of *concert of action* (his courtroom version of *conspiracy and coverup*). The story was unearthed but never displayed for the public to view.

Until now.

# ADDITIONAL RESOURCE for ALUMINUM WIRING INFORMATION

Visit **www.aluminumwire.info** for:

- A Report for Homeowners: ***Reducing the Fire Hazard in Aluminum-Wired Homes***

- A Report for Fire Investigators: ***Fire Due to Overheating Aluminum-Wired Connections***

- Technical articles, papers, and reports

- Photographs

- Links to Beverly Hills Supper Club Information and Photographs

- Links to related websites

- Some of the documents cited in this book

And more.

# ABOUT THE AUTHOR

DR. JESSE ARONSTEIN IS WELL KNOWN for his work on two residential electrical fire safety problems: Aluminum Wiring and Federal Pacific Electric circuit breakers. His involvement started in the mid-1970s with work performed for the Consumer Product Safety Commission, and continued to the present. He has performed fundamental research on aluminum connection failure mechanisms and has investigated and tested the ways that fire ignition can result. He has authored more than 20 technical papers on his research and testing of the properties, behavior, and failure mechanisms of aluminum wire connections.

He holds bachelor's and master's degrees in mechanical engineering, and a doctorate in Materials Science. His industrial experience includes engineering positions at General Electric Co. (1957-1961), and both engineering and management positions at IBM Corporation (1961-1974) and Wright-Malta Corporation (1974-1984).

Since 1984, Dr. Aronstein has been a consulting engineer with a focus on electrical fire safety. He was deeply involved in many of the episodes recounted in this book and testified as an expert in the Beverly Hills Supper Club aluminum wiring trials and other related product liability lawsuits.

Dr. Aronstein is a Life Senior Member of IEEE, and has also been a member of ASTM, IAEI, and NFPA. He is a licensed Professional Engineer in New York State. He has 15 patents in his name and received several invention and achievement awards for his work at IBM.

# REFERENCES, SOURCES, and NOTES

THIS BOOK IS WRITTEN FROM THE VANTAGE POINT OF A NON-INDUSTRY participant. Information on the main topics, other than that of the author's personal experience and knowledge base, derives from the sources listed in this section.

## Introduction
*A genuine crisis ...*: R.J. Schoerner (Southwire), to "Gentlemen" (attendees at UL Apr.15, 1971 meeting) April 17, 1971

## Chapter 1—The Beverly Hills Supper Club Fire
Lawson, Robert G., *Beverly Hills, The Anatomy of a Night Club Fire*, Ohio Univ. Press, Athens, OH, 1984, ISBN 0-8214-0728-7

Elliot, R., Inside The Beverly Hills Supper Club Fire, Turner Publishing Co., Paducah, KY, 1996, ISBN 1-56311-247-7

Best, Richard L., *Reconstruction of a Tragedy*, The Beverly Hills Supper Club Fire, NFPA No. LS-2, ISBN No. 0-87765-113-2

Note: the following quotes in Best are portions of KY State Police interview transcripts

*I had turned around ...*: Best, p. 20
*We had just sat down. ...*: Best, p. 21
*Someone said something ...*: Best, p. 32
*And these comedians ...*: Best, p. 34
*And we quickly ...*: Best, p. 36
*I wasn't thinking about ...*: Best, p. 39
*At the front of the club ...*: Best, p. 43
*... there was no question ...*: Best, p. 38
*At about 11:30 p.m. ...*: Best, p. 49

Photo: J. Aronstein

## Chapter 2—Fatal Fire—Hampton Bays, New York

*On April 28, 1974, two people died* ...: Publication #516, Repairing Aluminum Wiring, U.S. Consumer Product Safety Commission, June 2011

*On April 28, 1974, in the early morning* ...: Testimony of R. Hersh: *Hazard Posed By "Old Technology" Aluminum Wiring Systems*, Hearings before the Subcommittee on Oversight and Investigations, Committee on Interstate and Foreign Commerce, House of Representatives, Ninety-Fifth Congress, Second Session, June 2 and 5; Sept. 25, and October 11, 1978, Serial No. 95-143, U.S. Government Printing Office, Washington, D.C., 1978

Details of the Hersch fire: K.Jones, Southampton Fire Department Report, April 4, 1974

"Investigators from the CPSC and ...": S. Greenwald (NBS) to J. Rabinow (NBS), May 14, 1974

"Responding to a request for information ...": L.Fisher (NY Consumer Protection) to H.M.Miller (NY Assemblyman), Dec.11, 1974

## Chapter 3—Kaiser Aluminum and Chemical Corp.

"The ongoing aluminum connection problem ...": C.E. Baugh, *Better Connector Life Vital to the Use of Aluminum in Distribution*, Electrical World, June 2, 1952, also as conference paper, Edison Electric Institute (EEI), May 7, 1952; Minutes, 58th meeting of Transmission and Distribution Committee of EEI, Chicago IL, May 6-7, 1952

"When UL bowed to the pressure, ...": A.Small (UL) to M.M. Brandon (UL), with attached press release, Aug. 26, 1946

"Almost immediately, U.S. Rubber ...": P.W. Baker (US Rubber) to UL, July 7, 1946

# REFERENCES, SOURCES, AND NOTES 231

"UL did not budge from its position ...": M.M. Brandon (UL) to P.W. Baker (US Rubber), August 2, 1946

"The 1950 US Rubber catalog of ...": US Rubber Co., Wire and Cable Catalog, 1950

"In 1951 UL accepted aluminum building wires ...": H.I. Jehan (UL) to file Mar. 30, 1951

"Across the industry this procedure ...": H.A. Schilling (Kaiser), Project YX/100/14.3, Progress Report No. 1 (Final), April 1, 1955, p.26, 33

"Until 1965, when Kaiser introduced its KA-FLEX ...": Kaiser, *Insulated Aluminum Conductors - Design and Installation Manual*, Fifth Edition, 1964, p. 44

"In 1952, Kaiser embarked on a program ...": E.W. Greenfield (Kaiser) to P.P. Zeigler (Kaiser), July 28, 1952

"Kaiser also established a 'Connector Propaganda Comm ...'": G.N. Houck (Kaiser) to J.E. Menz (Kaiser), Oct. 24, 1952

"Eight years later, in ...": C.G. Sorflaten (Kaiser), *Preliminary Report on the Utilization of Aluminum SEU Cable in Present Meter Sockets, Service Panels, and Range Receptacles*, July 3, 1957

"His warning was firmly supported ...": C.G. Sorflaten (Kaiser) to P. Persons (Kaiser), Oct. 3, 1957

'The Bristol plant and its ...": A.L. Kolb (Kaiser) to P.J. Smith (Kaiser), August 2, 1960

*We are going to use Howard as* ...: G.N. Houck (Kaiser) to J.P. Moran (Kaiser), Mar. 29, 1956

"At about the same time, Kaiser's C.G. ...": C.G. Sorflaten (Kaiser), Report on Building Wire Terminations, April, 1958

"Very few of the manufacturers that he approached ...": C.G. Sorflaten (Kaiser), Sales contact report, April 25, 1958

"They had to discourage the sale ...": H.H. Weber (Kaiser) to W.K. Priestly (Kaiser), Sept.13, 1957

"As an example, Sorflatten tested range ...": C.G. Sorflaten (Kaiser) to D.C. Keenan (Kaiser), May 8, 1958

"taking a calculated risk": J.P. Moran (Kaiser) to C.W. Higbee (Kaiser), Sept. 19, 1957

"Sorflatten defended his rigorous approach ...": C.G. Sorflaten (Kaiser) to J.P. Moran (Kaiser) Jan. 14, 1958

"Eventually, connector testing and research ...": ACAP Committee (Kaiser), Minutes, of meeting Jan 30-31, 1958

"The first units constructed were scheduled to be wired ...": C.C. Hrenek (Kaiser) to F.C. Jannsen (Kaiser), Apr. 1, 1958

"Noting that there was much misunderstanding ...": H.B. Whitaker (UL) to manufacturers (various), Subject - Marking of Equipment to Indicate Suitability for Field Connection of Aluminum Conductors, Dec. 19, 1957

"UL provided a letter ...": H. Mapplethorpe, Jr. (UL) to P.H. Winter (Pass & Seymour), Aug. 11, 1958

"Kaiser's R.P. Brown visited the Ravenswood complex ...": R.P. Brown (Kaiser) to G.N. Houck (Kaiser) Aug. 9, 1963

"A few weeks later, Kaiser's H.H. Borup visited ...": H.H. Borup (Kaiser) to T.E. Bastis (Kaiser) Aug. 20, 1963

"In July, 1965, just after the company introduced ...": R.J. Webster (Kaiser) to J.B. Roche (Kaiser) July 7, 1965, with copy of Kaiser advertisement, IAEI News, July, 1965

"In 1979, R.T. Noonan ...": R.T. Noonan (CPSC), *Affidavit ... Regarding Investigation at Ravenswood WV*, Aug. 22, 1979

"With the help of covert funding ...": The National Security Archive, https://nsarchive2.gwu.edu/news/20040925/index.htm

## Chapter 4—Fire, Columbus, Georgia

Johansen Fire, Description of events and witness quotes: Columbus GA fire department - transcript of taped firefighter and witness interviews;

REFERENCES, SOURCES, AND NOTES 233

Summary of Coroner's report; Report, *Fire at 5303 Grady Drive, Columbus GA*, U.S. Consumer Product Safety Commission, January, 1979; Affidavit, Robert J. Kelly, Aug. 29, 1979 (for CPSC)

*A genuine crisis* ...: R.J. Schoerner (Southwire), to "Gentlemen" (attendees at UL Apr.15, 1971 meeting) April 17, 1971

"In mid-1966, just a little more than ...": G.W. Pennington, City of Dayton OH, Dept. of Service and Buildings, Electrical Bulletin #26, June 20, 1966

"Pullen reported to NEMA that the ...": R.G. Pullen (NEMA), Trip Report - Dayton OH, July 11, 1966

"In November 1966 NEMA asked UL to ...": J.J. Kark (NEMA), Minutes of Meeting Nov. 16, 1966,

"The survey returns showed seven ...": G.E. Schall, Jr (UL) to J.J. Kark (NEMA), Mar. 31, 1970

"A proposed second phase of ...": J.J. Kark (NEMA), Minutes of Jan. 28, 1970 meeting, Ad Hoc Subcommittee on Questionaires

"... Leviton's chief engineer, wrote ....": M.J. Weitzman (Leviton) to K.H. Zimmerman (P.E.), Mar. 6, 1973

## Chapter 5—"Shootout" at Travelers Hotel

"Three wiring device manufacturers had gone on record ...": Report of the NEMA General Engineering Committee Meeting, Wiring Device Section, March 24-25, 1971; J.W. Faulkner (Anaconda) to S. Bunish (Anaconda), April 13, 1971, Subject - *NEMA Building Wire and Cable Technical Committee Meeting 4/6/71*

"Dick Shaul of the NEMA staff reported ...": E.W. Roberts (GE), *Trip Report ... Meeting ... NEMA HQ 10/29/70*, Nov. 6, 1970

"Half a year later, W.A. Farquhar ...": W.A. Farquhar (UL) to manufacturers (wire & cabble, wiring devices) April 12, 1971

"For several years prior to Farquhar's urgent ...": G.E. Leneus (Southwire),

*Typical Tests of Triple-E Aluminum Alloy Conductors with U.S. Household Receptacles*, Feb. 1, 1971

"Kaiser's S.G. Roberts wrote that ...": S.G. Roberts (Kaiser) to E.J. Westerman (Kaiser), April 15, 1971

*The meeting was presided over by Mr. W.A. Farquhar ...*: W.H. Roemer (Cerro) to C.M. McCormack (Cerro), April 16. 1971

*I went to the meeting expecting it to ...*:.B.Hondalus (Reynolds), contact report, UL April 15, 1971 meeting, April 16, 1971

"Additional details ...": D.S. Medrick (Anaconda) to H.C. Witthoft, report of UL April 15 meeting, April 19, 1971

"A few days later, ...": J.W. Mitchell (Essex) to W.A. Farquhar (UL), April 20, 1971

"For single-family houses ...": Affidavit to CPSC, R. Thompson, Sept. 7, 1977

## Chapter 6—A Fire Chief's Perspective

*... Gentlemen, the termination of aluminum ...*: Transcript of Duncan presentation, sent by C.A. Parris (NEMA) to Member Companies, Wiring Device Sec., April 26, 1971

"Duncan later appeared in 1978 ...": *Hazard Posed By "Old Technology" Aluminum Wiring Systems*, Hearings before the Subcommittee on Oversight and Investigations of the Committee on Interstate and Foreign Commerce, House of Representatives, Ninety-Fifth Congress, Second Session, June 2 and 5; Sept. 25, and October 11, 1978, Serial No. 95-143, U.S. Government Printing Office, Washington, DC, 1978

## Chapter 7—Underwriters Laboratories

*From the outset ...*: T.A. Edison (The Edison Electric Light Co.) to NY Board of Fire Underwriters, May 6, 1881

"Matson wrote to Small expressing concern ..." A.F. Matson (UL) to A. Small (UL), May 1, 1946

# REFERENCES, SOURCES, AND NOTES 235

"... definite fire hazard ...": A. Small (UL) to M.M. Brandon (UL), May 9, 1946

"In July 1946, the U.S. Department of Agriculture ...": J.R. Cobb (USDA) to UL, July 31, 1946

"Brandon responded for UL ...": M.M. Brandon (UL) to J.R. Cobb (USDA), Aug. 2, 1946

"Brandon questioned the aluminum ...": M.M. Brandon (UL) to A. Small (UL), Aug. 2, 1946

"UL published a summary of the test program ...": *Aluminum Building Wires and Connectors*, Bulletin of Research #48, UL, Sept. 1954

... *Although he was aware of the* ...": *Excerpts From Trip Reports of Messrs. W.C. Schwan and R.G. Pullen - 1969-70 - Aluminum Terminations*, NEMA, Feb. 10, 1970 (last entry date)

... *Recently the Laboratories has been advised* ...: E.J. Coffey (UL) to R. Rowland (Southwire), Jan. 18, 1971

*It is recognized that, as an independent* ...: UL Listing and Follow-up Services Agreement Contract, Cadillac Cable Corp., April 7, 1969 (as an example)

"H.J. Kontje, a former UL Executive ...": Deposition, H.J. Kontje, Manoma Realty Management LLC v Federal Pacific Electric Co., U.S. District Court, Southern District of New York

... *I do recall our midnight safari* ...: H.M. Dreher (NEMA) to R.S. Keith (Kaiser), Nov. 19, 1957

"Later, in 1972, when Kaiser's K-140 ...": R.S. Keith (Kaiser) to A.S. Hutchcraft (Kaiser), May 19, 1972

"The wire that Kaiser submitted for the program as K-140 ...": L.A. Kirkpatrick (Alcan) to D.A. Jeanotte (IBM), re: ASTM tests, Oct. 13, 1986

... *The recent enthusiasm for use of aluminum wire* ...: K.S. Geiges (UL) to H.B. Whitaker (UL), Mar. 31, 1959

*... I believe your suggestion to include some information ...*: H.B. Whitaker (UL) to K.S. Geiges (UL), Apr. 14, 1959

## Chapter 8—Mobile Home Fires

Photos: Belden Family photos salvaged after the fire

*I went into my girls' room* ...: Janet Belden, ABC TV *20/20* segment "Hot Wire", April 3, 1980

"Mobile homes account for ...": www.bbc.com/news/magazine-24135022

"The manufacturers saved an estimated ...": R.S. Keith (Kaiser) to R.L. Shelton (Kaiser) Jan. 20, 1971; C.R. St.John (Kaiser) to A.S. Hutchcraft (Kaiser), May 5, 1972

"In 1966, the year after aluminum NM cable appeared ...": C. Stone (Kaiser) to D. Macnaught (Kaiser), Feb. 11, 1966

"In March of 1969 ...": A.G. Vande Weile (Mobile Home Owner) to H. McNee (Seattle, Chief Deputy Fire Marshall), Mar. 1, 1969 (date apporox., see attachments to Bastedo to Pickens March 17, 1969)

*The attached letters are frightening.* ...: E.H. Bastedo (Kaiser) to C.R. Pickens (Kaiser), A.G. VandeWiele, March 17, 1969 (with attachments)

"Kaiser's R.S. Keith wrote in his meeting report ...": R.S. Keith (Kaiser) to R.L. Shelton (Kaiser) Jan. 20, 1971

*They have had a considerable number of fires in Mobile Homes* ...: J. P. Moran (Aluminum Assoc.), Meeting with Dell, Mar. 17, 1971

"Many of the connections overheated immediately ...": O.G. Wedekind (UL) to E.J. Coffey (UL), July 7, 1970

"In spite of all the precautions and retightening ...": O.G. Wedekind (UL) to W.A. Farquhar (UL), Dec. 10, 1975

*Attached is a tabulation of* ...: O.G. Wedekind (UL) to W.A. Farquhar (UL), May 4, 1971

"In an internal May, 1972 memo ...": C.R. St. John (Kaiser) to A.S. Hutchcraft (Kaiser), May 5, 1972, re: visit to Melody Homes

REFERENCES, SOURCES, AND NOTES 237

"Later that year, Foremost Insurance Company ...": J.D. Hosey (Foremost Insurance Co.,) to M Weitzman (Leviton). Jan. 7, 1972, with attachments

"American Plan, another insurance company ...": Wall St Journal article, *Insuring Mobile Homes Wired With Aluminum*, Sept. 22, 1972

"In 1973 a bill was introduced ...": *Congress Considers Mobile Home Legislation*, Rural Electrification (magazine), May, 1973, p. 20

"A local professional engineer ...": F.C. Jones (P.E., Consultant) to W.A. Farquhar, Feb. 17, 1973

"The UL and industry people ...": O.G. Wedekind (UL) to W.H. Farrell (UL) Feb. 16, 1973

"The 1971 Edition ...": National Electrical Code, National Fire Protection Assoc., 1971 edition, Section 110-14

"At a previous hearing ...": *Aluminum wiring in homes called dangerous*, The Arizona Republic (newspaper), Feb. 9, 1973, p. 23

"Kaiser's representative at the hearing, R.L. Shelton ...": R.L. Shelton (Kaiser) to P. Mara and H.Cook (Aluminum Assoc.), Feb. 20, 1973, with attachments

"At about 8:00 in the morning ...": R.J. Kelly (CPSC), Affidavit, Aug. 29, 1979, with attached exhibits

## Chapter 9—Trial By Television

"A full-page Kaiser advertisement ...": Kaiser advertisement "Trial by Television" (various major newspapers), April 1980

*In effect we are all Technical Marketeers* ...: R.S. Keith, "Technical Sales Handles", Kaiser, Feb. 8, 1967

"In January of the same year ...": Transcript, *Byline - Leah Thompson*, WRC-TV (NBC affiliate) , Jan. 26, 1980

"Kaiser was not the first company ...": W.H.Miller, *Fighting TV Hatchet Jobs*, Industry Week, Jan 12, 1981 p.61

*The capitulation, advised by network lawyers* ...: P. J. Boyer, "Kaiser Case

May Hamper TV News Probers", Associated Press, as in Schenectady Gazette, Nov.27, 1980, p. 54

*... These statements, as well as the program ...*: Article, *Kaiser Files FCC Complaint Against ABC*, NY Times, Feb. 13, 1981

*Roon Arledge, President of ABC News ...:* John O'Connor, *TV View - When Critics of TV News Get to Talk Back*, NY Times, Aug. 2, 1981

"Flach did not mention ...": G.W. Flach (City of New Orleans) to J. Aronstein (Consulting Engineer), June 22, 1982

## Chapter 10—Fatal Fire, Houston Texas, 1978

"The senior partner of the firm ...": M. Curriden, *Richard "Racehorse" Haynes*, ABA Journal (online), March 2, 2009, https://www.abajournal.com/magazine/article/richard_racehorse_haynes

"This could account for ...": J.R. Schultz (Kaiser) to W.A. Wong (Kaiser), May 22, 1974; J. Aronstein, *Test for Loosening of "Old Technology" Alum Conductor in Binding Head Screws*, CPSC Task V Report, Contract C-79-0079, Wright-Malta Corp., Feb. 10, 1981; J. Aronstein, Loosening of Aluminum Wired Binding Head Screw Connections on Canadian Receptacles, Wright-Malta Corp., June 18, 1980;

"Next, engineers at Kaiser demonstrated that ...": T.R. Pritchett (Kaiser) to W.H. King (CPSC), July 19, 1974

"Lastly, and perhaps most common ...": O.G. Wedekind (UL) to E.J. Coffey (UL), July 7, 1970

"Williamson coined the phrase ...": J.B.P. Williamson, *The Self-Healing Effect - Its Implications in Accel Testing of Conductors*, Proceedings of the 10th International Conference on Electrical Contacts (ICEC), Budapest, Hungary, August 1980

## Chapter 11—The UL Ad Hoc Committee

*In response to Baron Whitaker's letter ...*: W.H. Abbott (Battelle) to W.A. Farquhar (UL), Aug. 30, 1971

# REFERENCES, SOURCES, AND NOTES

"Eventually, Whitaker and Farquhar convened ...": W.A. Farquhar (UL) to E. Buja (Anaconda), June 1, 1972

"They had previously agreed ...": H.P. Michener (NEMA) to W.A. Farquhar (UL), July 14, 1972

"According to Ed Buja ...": E. Buja (Anaconda) to H.C. Witthoft (Anaconda), Aug. 8, 1972

"The attendees agreed ...": H.B. Whitaker (UL) to members of Ad Hoc Committee, Aug. 3, 1972

"Buja again attended ...": E. Buja (Anaconda) to H.C. Witthoft (Anaconda), Aug. 23, 1972

"Hatfield Wire and Cable halted production ...": T.J. Stewart (Hatfield) to All Agents and District Sales Offices, Oct. 30, 1972

"Battelle's W. H. Abbott challenged, among other things ...": W.H. Abbott (Battelle) to W.A. Farquhar (UL) Oct. 4, 1972

"Buja, in his usual meeting report ...": E. Buja (Anaconda) to H.C. Witthoft (Anaconda), Nov. 21, 1972

"Whitaker, in his report (minutes) of that meeting ...": H.B. Whitaker (UL) to members of Ad Hoc Comm., Nov. 17, 1972

"Whitaker had written early on ...": H.B. Whitaker (UL) to members of Ad Hoc Comm., Aug. 23, 1972

"J.F. Tibolla was chairman of ...": J.F. Tibolla (Circle F) to H.C. Smith (NEMA), Nov. 28, 1972

"Adding to the objections, R.O. Wiley ...": R.O. Wiley (Bryant) to H. Leviton (Leviton), Nov. 30, 1972

"It was not to be. B. Falk, the NEMA President ...": H. Leviton (Leviton) to M.J. Weitzman (Leviton) Nov. 24, 1972

"Leviton, President of one of largest ...": H. Leviton (Leviton) to M.J. Weitzman (Leviton) Nov. 24, 1972

"Only a year earlier, UL sent out ....:" UL Bulletin, Nov. 8, 1971

"Several days after the meeting, Farquhar sent a letter ...": W.A. Farquhar (UL) to M.G. Domsitz (NBS) Nov. 20, 1972

"... L.M. Kushner of NBS had sent ...": L.M. Kushner (NBS) to J.E. Moss (U.S. Congress), Oct. 19, 1972

"According to Whitaker's summary of the next meeting ...": H.B. Whitaker (UL) to members of AD Hoc Comm., Jan. 16, 1973

"... The compiled pigtailing data ...": *Tabular Report Underwriters' Laboratories Survey: Aluminum Wire Terminations*, Underwriters Laboratories, Inc., Feb. 1973

"The results were discussed ...": S. Kolmorgen (Anaconda) to E. Buja (Anaconda), Mar. 1, 1973

"The UL pamphlet ...": *The Use of Aluminum Conductors With Wiring Devices in Electrical Wiring Systems*, Underwriters Laboratories, Inc., March 1973

"At the same time that the UL pamphlet ...": Transcript, CPSC Public Hearing on Aluminum Wire, April 18. 1974

"A 1996 field report ...": D. Jones (homeowner) to D. Friedman (Inspectapedia), email, May 5, 1996

"And another, from 1999 states ...": R.P. Zulli (Retired Electrician) to D. Friedman (Inspectapedia), email, Mar. 1, 1999

"Instead, he used stationary with the letterhead 'Ad Hoc Committee ...'": H.B. Whitaker (UL) to M. Smithman (Nat'l. Assoc. of Home Builders) May 22, 1973

*As a general statement this document* ...: W.H. Abbott (Battelle) to W.A. Farquhar (UL), Oct. 4, 1972

"To illustrate the point, in 1975, mobile home manufacturer Fleetwood ...": P. Hayes (Fleetwood) to C. Fitzpatrick (homeowner), Jan.31, 1975

"Whitaker then met with ...": H.B. Whitaker (UL) to members of Ad Hoc Comm., April 13, 1973

"NEMA President B. Falk wrote ...": B.H. Falk (NEMA) to H.P. Michener (NEMA), April 19, 1973

"Whitaker and the Aluminum Association also ...": H.B. Whitaker (UL) to members of Ad Hoc Comm., Oct. 31, 1973

REFERENCES, SOURCES, AND NOTES                                      241

## Chapter 12—The Consumer Product Safety Commission

*... It was about 6:00 P.M. on January 7, 1976 when the fire started ...*:
M.P. Barklay Affidavit to CPSC, May 5, 1976

*... I am familiar with general repair work ...*: R.R. Guindon, Affidavit to
CPSC, May 5, 1976

"In February 1972, about the same time that ...": B.H. Falk (NEMA) to
Specified Voting Representatives, Feb. 1, 1972

"The reports that Abbott produced ...": W.H. Abbott (Battelle), 2nd
Newsletter - Aluminum Connector Group, Nov. 17, 1972

*I feel that in all these meetings we have ...*: F.L. Gelzheizer (Bryant) to
"Gentlemen" (NEMA Comm. Members), Feb. 15, 1972

"One week later, Gelzheizer wrote ...": F.L. Gelzheizer (Bryant) to
E. Boquist (Bryant), Feb. 24, 1972

*The final GEC version appears ...*: J.F. Tibolla (Circle F) to E. Laskin
(Circle F), Mar. 6, 1974

"The prepared statement that Falk read ...": B.H. Falk (NEMA) to CPSC,
prepared statement for hearing Mar. 28, 1974

"After reading the prepared statement, Falk concluded ...": observer's report
(Kaiser), CPSC Hearings Mar. 27-28, 1974

"Since at least 1966, NEMA field men had ...": *Excerpts From Trip Reports
of Messrs. W.C. Schwan and R.G. Pullen - 1969-70 - Aluminum
Terminations*, NEMA, Feb. 10, 1970 (last entry date)

"An industry observer at the hearings ...": observer's report (Kaiser), CPSC
Hearings Mar. 27-28, 1974

"About one month after he testified ...": F. Cerra, *Aluminum Home Wiring
Poses Dangerous Fire Threat*, L.I. Newsday, October, 1974

"As an example, a month before ...": J. Rabinow (NBS) to W.A. Farquhar
(UL), Feb. 19, 1974

"Farquhar responded by letter ...": W.A. Farquhar (UL) to J. Rabinow
(NBS) Mar. 8, 1974

"CPSC field investigators from the Los Angeles office ...": CPSC In-Depth Report, Oxnard CA Fire, May 7, 1976

*The examination of existing field data* ...: E.D. Bunten et al., *Hazard Assessmant of Aluminum Elec Wiring in Residential Use*, NBSIR 75-677, National Bureau of Standards, Dec., 1974

"A summary of the CPSC's collection ...": R. Newman, *Hazard Analysis of Aluminum Wiring*, CPSC, April, 1975

*... June 75 - outlet glowing with nothing plugged into it* ...: B. Callahan, Affidavit to CPSC, May 3, 1976

"In 1977 the CPSC conducted an in-home survey ...": R. Newman and W.H. King, *Pilot Study of Branch Circuit Wiring Systems in Montgomery County*, CPSC, Sept., 1977

"The FRI summary states ...": Executive Summary, *National Controlled Study of Relative Risk of Overheating of Aluminum Compared with Copper Wired Electrical Receptacles in Homes and Laboratory*, Franklin Research Center, April 20, 1979

"In August, 1975 the CPSC ...": Article, *Aluminum Wiring Called Major Fire Risk in Homes*, Los Angeles Times, Aug. 9, 1975

## Chapter 13—"If Properly Installed"

*Although we believe that present* ...: W.A. Farquhar (UL) to manufacturers (wire & cabble, wiring devices) April 12, 1971

*Tenants in twelve apartments ... complained* ...: R. Newman, *Hazard Analysis of Aluminum Wiring*, CPSC, April, 1975, p. 17

"UL's original letters of approval ...": H. Mapplethorpe, Jr. (UL) to P.H. Winter (Pass & Seymour), Aug. 11, 1958

"Kaiser then explicitly promoted the use ...": Kaiser advertisement, IAEI News, July, 1965

"UL's chief electrical engineer, Farquhar, tried to clear up ...": W.A. Farquhar (UL), *Aluminum Conductor Termination, A Controversy or a Misunderstanding - Which?* IAEI News, January, 1966

# REFERENCES, SOURCES, AND NOTES 243

*The article in the January, 1966 issue* ...: F.E. Devlin (Bryant) to R.D. Gormier (Bryant) April 15, 1966

"Illustrating the lack of clear, uniform ...": H. Mitchell (Arrow-Hart) to M.J. Weitzman (Leviton), Mar. 5, 1971

*I recently had an electrical fire in a receptacle* ...: E.L. Schueler (Homeowner) to H.B. Whitaker (UL), Jan. 9, 1976

*However ... the credibility of your organization* ...: F.G. McClellan, Jr. (Orange Co. CA Dir. Bldg. Safety) to W.A. Farquhar (UL), Mar. 2, 1972

"Kaiser's first instruction manual for ...": H.W. Biskeborn (Kaiser), *Manual for Interior Wiring*, Mar. 15, 1954

"Kaiser salesmen were directed ...": R.S. Keith (Kaiser) to Elec. Products Salesmen, April 30, 1965

"Kaiser engineer reported Scotchlock twist-on ...": R.P. Brown (Kaiser) to G.N. Houck (Kaiser), Aug. 9, 1963

*I don't have any sure way of checking* ...: R.S. Keith (Kaiser) to A.W. Kolle (Kaiser), Dec. 18, 1972

"The proposed UL test procedure ...": E.W. Krawiek (UL) to Elec. Council and Manufacturers, Sept. 29, 1972

"In 1971 UL's O.G. Wedekind ...": O.G. Wedekind (UL) to W.A. Farquhar (UL), May 4, 1971

*... the design of the conventional* ...: O.G. Wedekind (UL) to E.J. Coffey (UL) July 7, 1970

"Farquhar generally relied on anecdotal hearsay ...": W.A. Farquhar (UL) to R.E. Ward (Chief Elec. Eng., Dept. Insurance & Banking, TN Div. of Fire Prevention) Mar. 24, 1971

"Farquhar claimed that the reason ...": ibid.

"... assumption that early failures must ...": R.S. Kieth (Kaiser) to J.W. Loveland (Kaiser), Aug. 2, 1972

"When one of Leviton's new technology ...": T.J. D'Agostino (UL) to W.A. Farquhar (UL) June 8, 1972

"... commented on a UL meeting report ...": C.S. Kolaczyk (UL) to Elec. Council and manufacturers, June 18, 1971

"Weitzman countered that UL should ...": M.J. Weitzman (Leviton) to C.S. Kolaczyk (UL), July 1, 1971

"Leviton's patent for its improved terminal ...": US Patent 3,944,314

"Responding to questions posed by Congressman Moss ...": J.N. Pearse (Leviton) to J.E. Moss (U.S. Congress), July 14, 1978

"For example, in 1991, an electrical specialist ...": R.H. Murray (NFPA/NEC) to R.L. Dougherty (The Colonie, condo Assoc.), May 30, 1991

"A 2006 article in IAEI News ...": C. Hunter, *Still Living in the 60's - Alum Building Wire and Terminations*, IAEI News, Jan.-Feb., 2006

"... as Leviton VP-Engineering M.Weitzman wrote ...", M.J. Weitzman (Leviton) to K.H. Zimmerman (P.E.), Mar. 6, 1973

"A dictionary definition of 'workmanship' is ...": https://dictionary.cambridge.org/us/dictionary/english/workmanship

## Chapter 14—Fatal Fire, Phoenix Arizona

*... Looks like one of our operators had a bad night ...*: T.C. Brown (Kaiser) to R.L. Shelton (Kaiser), April 5, 1967, with attached claim form

"Arthur Electric, an installer located in Phoenix ...": Claim #LA8-5B68-SL (Kaiser), Aug. 30, 1968; Claim #SL-73-006 (Kaiser), April 12, 1973

*Discovered that refrigerator was not working ...*: J.C. Williams (homeowner) Affidavit to CPSC, May 5, 1976

"San Diego, California is another place ...": Claim #LA72-028-SAN (Kaiser) July 31, 1972

"Three years later, in that same city ...": P.V. Mara (Aluminum Association) to J.J. Cashel (Kaiser) April 17, 1975

*A fire occurred in the crawl space ...*: Fire, Thousand Oaks CA, In-Depth Investigation Report (CPSC), Fire Mar. 10, 1979 (date of fire)

POSTSCRIPT 245

## Chapter 15—Canada

"The Toronto Housing Authority noted ...": 1975 Annual Report, Toronto Housing Authority

*If your house has aluminum wiring* ...: Article, *Aluminum Safe as Home Wiring*, Toronto Star, Nov. 26, 1975

[In the USA] *UL standards are written* ...: R.O. Wiley (Bryant) to G.R. Dunbar (Bryant), Feb. 17, 1977

*The Approvals Council (Electrical)* ...: Article, *New Receptacle Requirements for Aluminum Wiring*, (CSA), Standards Canada, v.6, No.3, Nov. 1975

*I notice you held a meeting on this subject* ...: J.G.C. Henderson (Canada Wire and Cable) to E.J. Coffey (UL), Mar. 18, 1970

"Canadian responses to one of the UL surveys ...": W.A. Farquhar (UL) to Chief Elec. Inspectors, Dec. 26, 1972

"In 1968, three years after Kaiser introduced ...": J. Lukaszewicz (ALCAN) to A.S. Lutgens (UL), Aug. 14, 1968

"The engineer assigned to respond was instructed ...": E.J. Coffey (UL) to R.F. Gloisten (with marginal note)

"In late 1970, at a NEMA meeting, CSA's ...": E.W. Roberts (GE) Trip Report, Nov. 6, 1970

... *I believe that in five years* ...: P. Haskelson (Leviton Canada) to A.A. Wanlass (CSA), Aug 5, 1970

"These receptacles remain installed in many Canadian ...": Internet postings, for example https://www.electrical-contractor.net/forums/ubbthreads.php/topics/124151/Hot_Receptacle.html (accessed Feb.2, 2021)

"The "special service" connectors on the market proved to be failure-prone ...": J. Aronstein and W.E. Campbell, *Overheating Failures of Aluminum-Wired Special Service Connectors*, IEEE Transactions, Vol. CHMT-6, No. 1, March 1983

"This new Canadian NM cable was also evaluated at W-M. ...": J. Aronstein, *Evaluation of an Aluminum Conductor Material for Branch Circuit Applications*, IEEE Transactions, Vol. CHMT-8, No. 1, March 1985

"The Wilson Commission submitted its report ...": J.T. Wilson, *Report of the Commission of Inquiry on Aluminum Wiring*, Queens Printer for Ontario, December 1978 (3 volumes), ISBN 0-7743-2863-0

*... It is unfortunate that, during the ...*: Wilson, ibid., Section 1.8.3

*The distinction between safety and reliability ...*: Wilson, ibid., v. 2, Section 2.1, last sentence

temperature, abnormal: *in functioning electric-wiring devices ...*: Wilson, ibid., v. 3, Section 3.2

"A homeowner group in Ottawa followed up ...": Article, *Irate homeowners sue for $375 million*, Montreal Gazette, April 24, 1979

*Temperatures of aluminum-wired receptacles ...*: J. Aronstein, *Summary Report - Study of Overheating of Aluminum-Wired Electrical Receptacles in Scarborough, Toronto, Homes*, Wright-Malta Corp., Jan. 11, 1982

## Chapter 16—An Industry On Trial

Judge H.R.Wilhoit Memorandum opinion re Beverly Hills Supper Club fire aluminum wire litigation: US District Court for the Eastern District of Kentucky - 639 F. Supp. 915 (E.D. Ky. 1986), May 13, 1986 https://law.justia.com/cases/federal/district-courts/FSupp/639/915/1745635/

"The company's promotional flyer ...": GE Promotional Flyer, *General Electric Presents - Aluminum Building Wire*, May, 1947

## Chapter 17—Arrested Development

*But, as with ...*: S.J. Gould, *Freudian Slip, Monthly column, "This* View of Life", Natural History, 31 Dec 1986, 96(2):14-21)

"The foundation for the new test ...": UL standard: UL486, *Wire Connectors and Soldering Lugs*, 3rd edition, Feb.1953 (typical)

"That is what led to Kaiser's adopting ...": see Chapter 3

"... and the UL-NEMA field failure survey ...:" see Chapter 7

"Its pass/fail performance requirements were weaker ...": J. Aronstein,

# REFERENCES, SOURCES, AND NOTES 247

*Qualification Criteria for Aluminum Connections*, Transactions of the 32nd IEEE Holm Conf. on Electrical Contact Phenomena, Chicago, IL, 1986

*During the discussion* ...: W.A. Farquhar (UL) to H. Cook (Aluminum Association) July 7, 1971

"A combination of stress tests and life tests ...": W. Yurkowsky et al, *Accelerated Testing Technology* (2 volumes), Hughes Aircraft, Feb 1966

"Inconsistent results were obtained ...": see Chapter 7

... *at 418 cycles* ...: R.A. Bates (Pass & Seymour) to W.R. Bowden, Jr. (Pass & Seymour), Aug. 13, 1973

"The surprise was that ten days of sitting idle ...": E.W. Krawiek (UL) to R.A. Bates (Pass & Seymour), Sept. 12, 1973

"A similar incident occurred at Kaiser ...": G.C. Wolfer (Kaiser) to W.A. Wong (Kaiser), Sept. 12, 1975

*Closely related to this work* ...: W.H. Abbott (Battelle), *Studies of Aluminum Wire and Cable Terminations at Battelle-Columbus*, undated summary regarding group-sponsored research program that was initiated June, 1971

*The next point we should consider* ...: G.H. Kissin (Kaiser) to M. Thelen, Jr. (Thelen, Marrin, et al), Dec. 22, 1975

"Another surprise occurred during UL's listing ...": T.J. D'Agostino (UL) to W.A. Farquhar (UL) June 8, 1972

... *Both the "old" Kaiser material* ...: E.W. Krawiec (UL), to E.J. Coffee (UL), March 13, 1973

"On the other hand, Farquhar's reported objective ...": see Chapter 5

*The primary objective* ...: T.J. D'Agostino (UL) to R.S. Keith (Kaiser), June 1, 1973

... *The UL people* ...: C.R. St. John (Kaiser) to T.W. Franklin and P.E. Mueller (Kaiser), July 3, 1972

"In about 1995, the Ideal #65 'Twister' ...": W.H. King, Jr. (CPSC) *Log of Meeting* June 1, 1995; A. Stadnik (CPSC) to J. Beyreis (UL), June

15, 1995; J. Killinger (UL) to A. Stadnik (CPSC), July 28, 1995; J. Aronstein (Consulting Engineer) to W.H. King, Jr., Oct. 10, 1995; B. Meier, Article, *Sparks Fly Over Industry Safety Test*, NY Times, Dec. 24, 1995; J. Aronstein (Consulting Engineer) to W.H. King, Jr., (CPSC) Jan. 3, 1996; J. Aronstein (Consulting Engineer) to T. Castino (UL), Jan. 4, 1996; J. Aronstein (Consulting Engineer) to W.H. King, Jr. (CPSC), Jan 24, 1996; W.H. King, Jr., (CPSC) to L.Mitchell (CSA), Jan. 31, 1996; L.S. Loeb (Landlord Resources) to W.H. King, Jr. (CPSC), Oct. 17, 1997; J. Aronstein (Consulting Engineer), *Evaluation of a Twist-on Connector for Aluminum Wire*, IEEE Holm Conference, 1997; J. Aronstein (Consulting Engineer) to T. Castino (UL), Dec. 16, 1997; L. Dosedlo (UL) to J. Aronstein (Consulting Engineer), Jan. 23, 1998; W.H. King, Jr. (CPSC) to D. Dini (UL), Sept. 4, 1998; S. Ratkovich (Ideal Industries Inc.) *Fact Sheet – Ideal #65 Twister*, 1998; J. Aronstein (Consulting Engineer), *Analysis of Field Failures of Al-Cu Pigtail Splices Made With Twist-On Connectors*, IEEE Holm Conference, Oct. 1999

"... but more than a decade passed before ...": UL-1567, *Receptacles & Switches for use with Aluminum Wire*, 1st Edition May 9, 1984; UL-486B, *Wire Connectors for Use With Alum Conductors*, 1st Edition, Mar 9, 1982; UL-486C, *Standard for Splicing Wire Connectors*, 1st Edition, Dec 30, 1983

"The AA-8000 series registration ...": P. Pollak (Aluminum Assoc.) to J. Aronstein (Consulting Engineer), Apr. 21, 1993

*We have Concluded* ...: C.R. St. John (Kaiser) to W.H. Cundiff (Kaiser), Jan 12, 1972

*Battelle believes* ...: W.H.Abbott, Battelle, to Aluminum Electrical Connector Group I, *Summaries of Research Results*, Battelle Columbus Laboratories, Jan 16, 1974

*There are entirely* ...: W.A. Farquhar (UL) to K.L. Bellamy (Ontario Hydro), Dec. 31, 1970

REFERENCES, SOURCES, AND NOTES          249

"Lighting problems led Chrysler ...": Chrysler LeBaron/Town & Country Owner's Manual, 1985, p.106

"ALCO and GM ran into costly aluminum connection failure ...": W. Cuisinier, *Alco's Unfortunate C415*, Rail Fan and Railroad, v. 5 No. 6, Sept. 1984; A.F. Koctur (GM Electro-Motive Division) to C.G. Sorflaten (Kaiser) July 10, 1959

"A report issued by the Canadian ...": *Reduction of Fires Caused by Residential Service Entrance Panel Boards*, Canadian Electrical Association, Report 134 U SOA, undated (approx. 1976)

*Summarizing, I think that* ...: J. Rabinow (NBS) to W.G. Leight (NBS), May 8, 1975

I am Fire Marshall for ...: P.M. McDonald, Affidavit to CPSC, Sept. 8, 1977

## Chapter 18—The Myth of the Self-Regulating Industry

*Historically, the fire death rate* ...: U.S. Fire Administration, *Fire Death Rate Trends: An International Perspective*, Topical Fire Report Series, v. 12, no. 8, July 2011

Residential fire statistics: R. Campbell, *Home Electrical Fires*, NFPA, March 2019 (online)

... *According to your* ...: M.M. Gilbert (DuPont), to W.H. Hoffman (UL), Dec. 11, 1970

... *As you may know,* ...: J.H. Watt (NFPA) to W.E. Bryant (MD Dept. of Economic and Community Dev.), Feb. 28, 1973

"The UL product listing report for ...": Listing Report, UL File E13399, p. T2-1a, issued July 14, 1972, new Sept. 15, 1975

"Since then, failure of aluminum-wired CO/ALR ...": J. Aronstein (Wright-Malta Corp., to W. Hamp (UL), Nov. 18, 1983; W.Hassan, *Investigation of Connection Failure of CO/ALR Receptacles*, Report #78-71-K, Ontario Hydro, Feb. 14, 1978; M.Leger, *Metallurgical Analysis of Failed COALR Devices*, Report #78-54-K, Ontario Hydro, Feb. 2, 1978; J. Aronstein (Wright-Malta) to W. Hamp (UL), May 25, 1983;

Wilson, ibid., v. 3, Section 3.1, Exhibit 71; E.O. Moore (Homeowner) to J. Aronstein (Consulting Engineer), Dec. 10, 2001; R.Jones, (Caine Company) to Engineering Dept. (Leviton Mfg. Co.), Mar. 6, 1981

*The mistake* ...: Announcer's closing remarks, *The Johnstown Flood*, "American Experience", PBS, Season 5, Episode 4, first aired Nov. 4, 1992

# INDEX

**Symbols**
*20/20* TV show *See* ABC and *20/20* show

**A**
AA-8000 Alloys 211
Abbott, W.H. 111, 114, 119, 128, 136–137, 212
ABC and *20/20* show 49, 77–78, 89, 91–95, 99–101, 192
abrasion
 costs of 54
 need for 33, 37, 54, 72, 159
 use in Ravenswood housing project 38
ALCAN
 about 177
 company town of Arvida 179
 IAEI News article 167
 media coverage and 98–99
 NUAL alloy wire 183
 request for information from UL 180
 Wilson Commission and 183
ALCO 213
Alcoa 57, 58, 189
AL-CU marking
 backwiring in Canada and 181

continuing confusion over 48, 128
 as "fix" 37, 47
 as gimmick 47, 167
Allende, Salvatore 39
ALUMICON 210
aluminum *See also* aluminum wire
 cost reductions from 213
 prices 39
 supply of 31
 WWII and 29, 31
Aluminum Alloy Assessment Program 207
Aluminum Association
 Arizona ban and 85
 coordination with UL 46, 85
 registration with 211
 UL ad hoc committee and 113, 114, 122, 129, 134
 UL crisis meeting (1971) and 54
aluminum oxide
 disinterest in testing 53
 formation of 33
 as fundamental problem 33, 53, 163, 165, 211–212
 tightening of screws and 108
aluminum wire *See also* bans on aluminum wiring
 Alloy Assessment (UL) 207

alloys, costs of 54
alloys, as "fix" 118, 183
alloys, testing 54, 74, 148–149, 208, 211
construction process 170
copper-clad aluminum wire 54, 64, 212
cost savings per residence 59, 78
decline in use 84, 151
EEE alloy 52, 56, 208, 211
fundamental problems with 33, 163–165
gains in market share 44
halts in production 93, 117
introduction of residential use 24, 31
K-140 alloy wire 74
numbers of homes with 143, 179
outdoor transmission lines 30
pressure for withdrawal of approval 49, 51, 111–112
process joints 170
research challenges 53, 70, 72–73
resources on 226
sales of 32, 34, 35
separation of connector testing from wire testing 32

251

size associations with
    failure 81–82, 92, 162
  testing by UL 211
  UL permanent approval
    of 32
  UL temporary approval of 31
American Insulated 55
American Insurance
  Association 68
American Plan 84, 117
Anaconda Wire and Cable
  Co. 58, 113, 114, 125, 150,
    189
Arizona, ban proposal in 85–86
Arizona State Division of
  Building Codes 87
Arledge, Roon 101
Arthur Electric 172
Arvida, Quebec 179
auto industry
  aluminum base bulbs 213
  safety in 218
aviation safety 217

## B

backwiring 154, 180–181
bans on aluminum wiring
  Arizona proposal 85–86
  in California 55, 61–65, 71
  in Colorado 51, 53, 55, 56
  increase in 46, 51, 55, 84,
    131
  litigation over 64
  in Ohio 44
  in Washington 51
Barclay fire 133
Barclay, M.P. 133

Barrett, F. 139
Bastedo, E.H. 79
Battelle
  aluminum oxide and 212
  call for withdrawal by 111
  on need for real-world
    testing 205
  research sponsorships
    at 136
  testing by 209
  UL ad hoc committee
    and 114, 115, 119
Béland, Bernard 182
Belden, Candy 77
Belden fire (Wittman, AZ) 77
Belden, Irwin 77
Belden, Janet 77
Belden, Shelly 77
Beverly Hills Supper Club fire
  events of 9–15
  fatalities 7, 18
  first trial 189
  investigations and
    reports 16, 19–20
  lawsuits 7, 19–20, 76, 92, 99,
    182, 189–199, 223–225
  origin and cause 19–20, 23,
    189, 193, 196
  second trial 194–195
  in *20/20* progam 92
Blake, H. 87
Boeing 218
Borup, H.H. 38
Boyer, Peter 100
Brand, E.A. 114
Brandon, M.M. 68, 69
Breheny, M. 138–139
Broley, M. Walker 176

Brown, R.P. 37
Bryant 121, 136, 155
Building Industry
  Association 64
Buja, Ed 113, 119
Buning fire 105–109
Burndy Corporation 106

## C

cadmium plating of terminal
  screws 48
California
  ban in 55, 61–65, 71
  CPSC investigations and
    hearings 126, 141–142
  fires from defective wire 172
California State Fire Chief's
  Association 63–64
Campbell, W.E. 144
Canada
  backwiring in 180–181
  concerns in 175–178
  homeowner
    organizations 176,
      186–187
  media coverage 98–99, 183
  numbers of homes with
    aluminum wiring 179
  Wilson
    Commission 183–185
Canada Wire and Cable 179
Canadian Broadcasting
  Company (CBC) 98–99
Canadian Electric Association
  (CEA) 177, 214

Canadian Standards Association
  (CSA)  98–99, 177, 180–181,
  183, 186
cause of fire
  Beverly Hills Supper Club
    fire  21, 23
  Buning fire  106
  defined  23, 178
  Hersh fire  23
  Johansen fire  43
CBC (Canadian Broadcasting
  Company)  98–99
CEA (Canadian Electric
  Association)  177, 214
Cerro  55, 57, 189
Chesley, Stanley  190, 191, 192,
  195–196, 223–226
Chicago World's Fair  67
Chile  39
Chrysler  213
CIA  39
Circle F Industries  121, 138
CO/ALR devices
  distribution of  208
  as "fix"  118–120
  testing of  148–149, 162,
    204, 206, 221
  UL ad hoc committee
    and  128
  wire wrap direction  161
Coffey, E.J.  71, 140
Colonial Village housing project
  (Ravenswood, WV)  36
Colorado, ban in  51, 53, 55, 56
Columbia  55, 57
Commission of Inquiry on
  Aluminum Wiring  183
concert of action  191, 223–225

Congress
  Consumer Product Safety
    Commission founding  132
  correspondence with
    industry  164, 165
  hearings  24, 63
  mobile home safety
    concerns  84
  *20/20* show and  96
  UL ad hoc committee
    and  113, 114, 123
connections *See also* receptacles; *See also* pigtailing
  AL-CU marking  37, 47, 48
  as cause in Beverly Hills
    Supper Club fire  21
  as cause in Hersh fire  23, 24
  as cause in Johansen fire  43
  failure rate  33
  failure research
    difficulties  30
  failures in outdoor use  30
  failures of splicing
    connectors  79
  Kaiser's awareness of
    problems with  33
  loosening of  82, 108
  as "self-healing"  109
  separation of connector
    testing from wire
    testing  32
  "special service"
    connectors  182
  testing by Dupont  220
  testing by Kaiser  34–36
  testing by UL  32, 209
Consumer Product Safety
  Act  132

Consumer Product Safety
  Commission (CPSC)
  on blaming poor
    installation  165
  California hearing  126
  on confusion over "proper"
    installation  153
  defective wire
    investigation  170,
    172–173
  *Hazard Analysis—Aluminum
    Wiring*  143, 186
  in-home survey  147
  investigations by  26, 43,
    132, 133, 139, 140–143,
    151, 214
  lawsuits against  105
  legal action on aluminum
    wire  21, 147, 150, 192
  pigtailing and  126
  Publication #516  150, 210
  Ravenswood housing project
    and  38
  testing by  144–145,
    148–149, 151, 223
  *20/20* show and  91, 93
  Twister and  210
  Washington, D.C.
    hearing  138–139
  Wilson Commission and  185
continuity tests  170
COPALUM connector  149, 210
copper
  copper-clad aluminum
    wire  54, 64, 212
  prices  31, 39, 48, 151
  supplies of  31, 69

wiring as standard for
    residences 24, 31
wiring practices 158
CPSC *See* Consumer Product
    Safety Commission (CPSC)
*CPSC v. Anaconda et al.* 150
creep 108, 164
Cromwell, William N. 27
CSA (Canadian Standards
    Association) 98–99, 177,
    180–181, 183, 186

### D
D'Agostino, T.J. 74, 76, 125,
    140, 207
daisy chain wiring 26
Dayton, OH ban 44
Dell, W.E. 80
Delta Airlines 218
Derus Media Services 130
discovery in lawsuits 69, 99,
    191, 196, 198
Domsitz, M.G. 114, 123
Downs, Hugh 91, 93
Dreher, H.M. 74
Druckman, Eileen 10
Duncan, Carl 61–65
Duncan meter terminal,
    testing 34
Dupont 220

### E
Eagle 189
EC grade wire
    KA-FLEX as 75, 208

as profitable 54
    UL ad hoc committee
        and 120
Edison Electric Institute 114
Edison, Thomas A. 67
EEE alloy wire 52, 56, 208, 211
electric utilities and UL ad hoc
    committee 114, 117, 129
Erskine, Brik 169, 173
Erskine fire 169, 173
Essex 58

### F
Falk, B. 122, 130, 138
Farquhar, W.A.
    ad hoc committee and 112–
        114, 119, 123
    on aluminum oxide 212
    article on installation
        instructions 155–158
    blame on poor
        installation 162–165
    meeting on crisis (1971) 44,
        52–55, 153, 162, 219
    responsibility for UL
        failures 49, 71–72
    suppression of letter by
        Gelzheizer 137
    testing by Wedekind and 83
    on testing program
        effectiveness 204
    on *20/20* 49, 93
fatalities
    Beverly Hills Supper Club
        fire 7, 18
    Buning fire 105
    from defective wires 172

fire death rate 217
Johansen fire 42
Federal Communication
    Commission (FCC) 99, 100
field surveys 45, 142–143,
    143, 202
firefighters
    Beverly Hills Supper Club
        fire 11, 15–17
    Carl Duncan on aluminum
        wire 61–65
    Hersh fire 25
    resolution by 63
fire hazard
    defining 178
    threshold of 146, 186
fire terms, defining 178
Flach, George 102
Fleetwood 128
Florida, bans in 84
Ford Motor Company 218
Foremost Insurance
    Company 84
Forum *See* UL ad hoc committee
Franklin Research Institute 147,
    149
Frei Montalva, Eduardo 39
Frey, L., Jr. 84

### G
Geiges. K.S. 75–226
Gelzheiser, F.L. 136–137
General Cable 57, 189
General Electric
    Beverly Hills Supper Club
        fire 189, 197–199
    Buning fire 106

call for withdrawal of
    approval  51
catalog  157
promotion of aluminum
    wire  198
General Motors (GM)  213
Georgia
    bans in  84
    lack of distribution of safety
        message in  131
Gilbert, M.M.  220
Gloisten, R.F.  158
glowing red-hot  145, 181, 203
GM  213
Gore, Albert  64–65
Gould, Stephen J.  201

## H

Harrell, Ronnie Larry  42
Haskelson, P.  181
Hatfield Wire and Cable  55, 117, 134
Haynes and Fullenweider  106
Haynes, Richard  106
*Hazard Analysis—Aluminum Wiring*  143–144, 186
hearings
    Canadian  183–185
    Congressional  63
    CPSC  126, 138
heat cycle testing  144, 201–202
heat, detecting in receptacles  25, 26, 78
Heighington, Gary  176, 187
Hensley  78

Hersh fire (Hampton Bays, NY)  23–25, 63, 132, 139, 145, 146, 223
Hersh, Jack  25
Hersh, Rosalind  24, 26, 63
Hersh, Sheri  25
Hewett, John  80
Hoffman, W.H.  220
homeowners
    experiences and Wilson Commission  184
    homeowner organizations  176, 186–187
Hondalus, B.  56
Horstman, R.L.  164
Houck, G.N.  35
humidity, effect on connections  82, 145, 205
Hutchcraft, A.S.  84, 102

## I

Iacocca, Lee  218
IAEI *See* International Association of Electrical Inspectors (IAEI)
IAEI News
    advertising by Kaiser  38
    blame on installation  167
    Farquhar letter on installation  155–158
Idaho
    concerns in  56
    pigtailing requirements  80
Ideal  210
inhibitors
    costs of  54

need for  33, 37, 54, 72, 159
    in splicing connectors  210
    use in Ravenswood housing project  38
inspections and inspectors
    concerns about mobile homes by  79–80
    free inspections in Canada  187
    of transmission line connectors  30
    pigtailing questionnaire and  123–124
    role of inspectors  124, 177
installation
    backwiring and  154, 181
    confusion over "proper" installation and  153–157
    by CSA  178
    defining proper installation  158
    lack of instructions and  115, 160
    in lawsuits  106
    loosening of connections and  108
    in media reports  102
    NEMA focus on  138
    UL ad hoc committee and  115, 118–119, 129
    UL crisis meeting (1971)  153, 161, 162–165
insurance industry
    refusal to insure mobile homes with aluminum wire  84, 117
    UL sponsorship  68

International Association of
  Electrical Inspectors (IAEI)
  IAEI News 38, 155–158, 167
  increased awareness of
    problem 51
  pigtailing questionnaire 124
investigations
  Belden fire 77
  Beverly Hills Supper Club
    fire 16, 19–20
  in Canada 183–185
  by CPSC 140–143, 151, 170,
    172
  of defective wire 170,
    172–173
  Hersh fire 26, 132, 139
  Johansen fire 43
  mobile home fires 141–142

## J

Jardell, Elizabeth 215
Jardell fire 215
Jardell, William 215
Jardell, William, II 215
Jerabek, R. 187
Johansen, Daniel 41–42, 49
Johansen fire (Columbus,
  GA) 41–42, 47, 49
Johansen, Janet 41–42
Johansen, Peggy 41–42, 49
Jones. F.C. 85
juror misconduct in Beverly Hills
  Supper Club trial 194

## K

K-140 alloy wire 74
KA-FLEX ALR wire 75, 142
KA-FLEX aluminum wire
  approval of despite test
    failures 208
  defective 168, 169–171
  Hersh fire and 27
  introduction of 33, 38, 40
  marketing of for residential
    use 28
  Ravenswood, WV housing
    project 36
  rebranding of 75
  sales to mobile homes 79
  testing of by Kaiser 28
Kaiser Aluminum and Chemical
  Corporation See also KA-FLEX
  aluminum wire
  aluminum oxide and 212
  Arizona ban proposal and 87
  awareness of connector
    problems 33, 79, 163
  awareness of mobile home
    problems 79, 84
  backwiring and 154
  Beverly Hills Supper Club
    lawsuit 189, 192, 193
  concerns about new UL
    specs 54
  halt of aluminum wire
    manufacture 93
  Hersh fire and 27–28
  installation instructions 33,
    37, 154, 159
  K-140 alloy wire 74
  Keith's role in 74
  lawsuit by CPSC 150, 192

lawsuits by victims 27–28,
    92, 94, 105
  marketing and public
    relations 34, 35, 38, 40,
    94–96
  origins and history of 29
  testing by 28, 34, 37, 50, 53,
    70, 98, 102, 202, 205
  *20/20* and 93–96, 99–101
  UL crisis meeting (1971)
    and 54, 57
  U.S. Rubber acquisition 34,
    74
Kaiser, Henry J. 29
Kaiser Permanente Metals
  Corporation 29 See
  also Kaiser Aluminum and
  Chemical Corporation
Keith, Roger S. 74, 80, 95, 160
King, Susan 97
King, W.H. 185
Kissin, G.H. 205
Kolmorgen, S. 125
Kontje, H.J. 73
Koppel, Ted 101, 102
Krawiec, E.W. 140, 206
Kushner, L.M. 123

## L

lawsuits
  and bans on aluminum
    wire 64
  Beverly Hills Supper Club 7,
    19, 76, 92, 99, 182,
    189–199, 223–225
  Buning fire 106

concert of action
  concept 191, 223–225
  against CPSC 105
  against CSA and Ontario
    Hydro 186
  documents from 69, 99, 191, 196, 198
  Erskine fire 173
  Hersh fire 27–28, 223
  Johansen fire 47
  media coverage and 98, 101
  20/20 show and 94
Leviton
  backwiring and 181
  Beverly Hills Supper Club
    lawsuit 189, 196
  blame on poor
    installation 165
  call for withdrawal of UL
    approval 50, 51, 55
  CO/ALR failures in
    testing 163, 206
  lawsuits and 47
  Oxnard, CA mobile home
    fire 142
  terminal screw testing
    and 48
  UL ad hoc committee
    and 114, 115, 122
  UL crisis meeting (1971)
    and 57, 58
Leviton, H. 122
Lloyd, Dick 52
Louisiana, bans in 84

## M

Malta Test Station 144–146 *See also* Wright-Malta Corp.
Manson, D.M. 180
marketing
  AL-CU marking as marketing
    gimmick 47, 167
  by Kaiser 34, 35, 38, 40, 74, 97
  role of UL in 46, 74, 221
Maryland, bans in 84
Matson, A.F. 68
McDonald, P.M. 214
McLellan. F.G. 157
media coverage
  in Canada 98–99, 183
  increase in 46
  20/20 show 49, 77–78, 89, 91–95, 99–101
  use of documents in Beverly
    Hills Supper Club trial 192
  Viewpoint 101–102
Medrick, D.S. 58
Melody Mobile Homes 84
Merrill, W.H. 67
meter bases, problems
  with 213
meter terminals, testing 34
Michener, H.P. 112
Mitchell, J.W. 58
mobile homes
  AL-CU connectors use 128
  fires in 77–79, 88, 133, 141–142, 215
  insurance and 84, 117
  numbers of 78
Moran, J.P. 36, 46, 80, 85
Moss, J.E. 123, 132, 165

Myers, S. 126

## N

Nader, Ralph 218
National Board of Fire
  Underwriters 68
National Bureau of Standards
  (NBS)
  Hersh fire investigation 26
  UL ad hoc committee
    and 114, 123
  UL testing methods and 140, 214
National Electrical Code (NEC)
  backwiring and 157
  compared to CSA 177
  on connections 86
  as political instrument 93
  reliance on UL
    testing 86–87, 116, 220
  20/20 show and 92
  UL ad hoc committee
    and 113, 126
National Electrical
  Manufacturers Association
  (NEMA)
  awareness of problems 138
  bans on aluminum wire
    and 44, 45
  Canadian concerns
    and 180–182
  coordination with UL and
    Aluminum Association 46
  CPSC hearings and 138
  field survey 202
  Gelzheiser letter and 137

General Engineering
  Committee  120, 138
  letter on call for action  135
  UL ad hoc committee
    and  112, 114, 120, 122,
    129
  UL crisis meeting (1971)
    and  54
  UL testing concerns  70
National Fire Protection
  Association (NFPA)  See
  also National Electrical Code
  (NEC)
  Beverly Hills Supper Club fire
    report  16
  blame on
    installation  165–166
  20/20 program and  92
NBS See National Bureau of
  Standards (NBS)
NEC See National Electrical
  Code (NEC)
NEMA See National Electrical
  Manufacturers Association
  (NEMA)
New York
  bans in  84
  lack of distribution of safety
    message in  131
  New York Board of Fire
    Underwriters  67
  New York State Director of
    Product Safety  27
NFPA See National Fire
  Protection Association (NFPA)
nickel plating  54
Noonan, R.T.  38
NUAL  183

**O**

odors, detecting overheating in
  receptacles  25, 26, 78
Ohio
  bans in  44
  lack of distribution of safety
    message in  131
Ontario Hydro  177, 183, 186
Oregon, pigtailing
  requirements  80
origin, area of
  Beverly Hills Supper Club
    fire  19, 23, 189, 193, 196
  Buning fire  106
  defined  23
  Hersh fire  23, 24, 26, 145
  Johansen fire  43
Oxnard, CA mobile home
  fire  141–142

**P**

pamphlet, UL ad hoc committee  117, 118–124, 156
Panama canal  28
Pass & Seymour  204
Pearse, J.  165
Pennington, G.W.  44
pigtailing
  in Canada  181
  with COPALUM
    connector  210
  costs of  80
  CPSC guidance and
    testing  145–146, 149, 150
  defined  79
  industry concerns over  117

inhibitors in  210
lack of testing on  119, 122
long-term concerns and
  tests  83, 118, 163
in NEC recommendations  86
problems with  119–124
questionnaire on  123–124
requirements for  79–80
with Ideal #65 Twister  210
UL ad hoc committee
  and  115, 119–124
Plaintiffs' Lead Council
  Committee (PLCC) in
  Beverly Hills Supper Club
  trial  190–199
plug-in/out operations, effect
  on connections  83, 149
Potter, F.  164
press release, UL ad hoc committee  117, 118, 129–130,
  134
Priestly, W.K.  36
Pritchett, T.R.  28, 163
process joints  170
public
  lack of involvement/
    warnings  73, 219
  UL's initial focus on  68
Publication #516 (CPSC)  150,
  210
public relations
  coordination between UL,
    NEMA, and Aluminum
    Association  46
  by Kaiser  34, 38, 40, 94–96
  UL ad hoc committee
    and  113, 116
Pullen, R.G.  44

PVC plastics  21

**R**
Rabinow, J.  140, 214
Ravenswood, WV housing project  36
receptacles
  AL-CU marking  37, 47, 48
  daisy chain wiring  26
  design issues  82
  detecting damage to  25, 26
  in Hersh fire  23, 24
  in-home survey on  147
  in Johansen fire  43
  listing of by UL without testing  24
  range/dryer  36, 37
  testing by CPSC  145
  testing by Kaiser  36, 37
  testing by UL  24, 70, 81–82, 86, 108, 161, 163, 204, 206
research *See also* testing
  challenges of  53, 70, 72–73
  secrecy of  136
  shift away from by industry  107, 136
  by Williamson  106
Research Bulletin #48 (UL)  69, 180
resources on aluminum wiring  226
Reynolds  56, 106
Rivera, Geraldo  49, 91–93, 99, 102–103
Rivington Apartments fire (Houston, TX)  105–109
Roberts, E.W.  51

Roberts, S.G.  54
Rodale  51, 164
Roemer, W.H.  55
Ross, J.  92
Rubin, Carl B.  189, 192
Rubin, Maury J.  27
Rural Electrification Administration  69

**S**
"safety message" press release  117, 118, 129–130, 134
St. John, C.R.  84, 209, 212
Schoem, Alan  165
Schoerner, Roger J.  8, 50, 56, 115
Schwan, W.C.  70
Scotchlock connectors  160
"self-healing" of connections  109
self-regulation myth of  219–226
service entrance cables
  defined  34
  in Kaiser tests  34
service panels, problems with  213
settlements
  Beverly Hills Supper Club fire  92, 193, 196–198
  Hersh fire and  28
Shaul, Dick  51
Shelton, R.I.  87
Shilling, Rick  11
Slater  189
Small, Alvah  31, 68
Smith and Stone  181

smoke, in Beverly Hills Supper Club fire  12–14, 20
Sorflatten, C.G.  34, 35–36, 37, 50
Southgate Fire Department (KY)  11
Southwire
  Beverly Hills Supper Club lawsuit  189
  EEE alloy  52, 56, 208, 211
  testing by  54, 211
  UL ad hoc committee and  115
  UL crisis meeting (1971) and  56
  warning by Schoerner  43
sovereign immunity  20
"special service" connector  182
splicing *See* pigtailing
Srivastava, K.D.  185
"Standards Canada" flyer  178
Stewart, T.J.  117–118
stress relaxation  108
stuffer, UL ad hoc committee  117, 118, 129
Sullivan and Cromwell  27–28

**T**
temperature
  effect on connections  82, 145
  fire hazard threshold and  186
  of connections in homes, survey  147
Tennessee, bans in  84
terminal screws

Buning fire and  106
causes of loosening  108
need for retightening  82, 83, 161
size of  82
stripping of  108
switch from brass to steel  48, 72
testing by CPSC  145
testing by UL  48, 82
wrapping wire on  161
testing  *See also* UL, testing by
alloy wire  54, 148–149, 208
by Battelle  209
CO/ALR receptacles  204, 221
continuity testing  170
copper-clad aluminum wire  54
by CPSC  144–145, 151
CPSC hearings and  139
by Dupont  220
funding of  70–71
heat cycle testing  144, 201–202
industry concerns about UL testing  34, 36, 70
by Kaiser  28, 34, 37, 50, 53, 70, 98, 102, 202, 205
lack of long-term testing  202–203
lack of real-world conditions in  82, 203, 205
lack of testing by UL  24, 37, 48, 53, 58–59
lack of testing of pigtailing  119, 122
secrecy around  53, 136, 139

separation of connector testing from wire testing  32
severity of failures and  203
by Southwire  54, 211
"special service" connectors  182
splicing connectors  79
terminal screws  48
thermal soak test  57
UL approval despite failing tests  69, 74, 162, 198, 206, 207, 221
by Wright-Malta Corp  144–146, 182, 183, 223
Texas
bans in  84
lack of distribution of safety message in  131
Texas Instruments (TI)  64, 212
thermal expansion  164
thermal ratcheting  108
thermal runaway  145, 149
thermal soak test  57
Thompson, Lea  97–98
Thorne, E.R.  87
Tibolla, J.F.  120–121, 138
tin plating  54
Toronto Housing Authority  175–176
Travelers Hotel meeting (1971)  43, 52–55, 153, 162, 219
Trimmer, Joyce  176, 178, 187
*20/20*  49, 77–78, 89, 91–95, 99–101
Twister (Ideal #65)  210

## U

UL ad hoc committee
first meeting  112–113
pamphlet  117, 118–124, 156
press release  117, 118, 129–130, 134
stuffer  117, 118, 129
thwarting of  109
UL, testing by
alloy assessment program  74, 207
approach as limited  202, 214
approval despite testing failures  69, 74, 162, 206, 207, 221
early program  32, 69
funding for  70–71
heat cycle tests  144, 201–202
industry concerns about  34, 36, 70
industry reliance on  86–87, 108, 116, 220
inhibitors and abrasion  159
inspector concerns about  44
later program  24, 70–71, 202–204, 209
misrepresentation of testing  83
secrecy and  140
separation of connectors and wires in  32
terminal screws  48
by Wedekind  81–82, 86, 108, 161, 163
wire  211

UL (Underwriters Laboratories, Inc.) See also UL, testing by; See also UL ad hoc committee
  access to industry research 136
  approval of aluminum wiring as permanent 32
  approval of aluminum wiring as temporary 31, 198
  backwiring and 154
  Beverly Hills Supper Club fire 21, 92, 191, 193
  blame on installer 153, 162–165
  Canadian requests for information and 180
  coordination with NEMA and Aluminum Association 46, 85
  denial of repsonsibility 116
  failure to live up to public expectations 46, 49, 58–59, 62, 71–74
  Farquar letter on installation 155–158
  Farquhar and letter by Gelzheizer 137
  field survey 45, 202
  focus on industry 69, 71, 73–75
  focus on public safety 68
  for-profit model 68
  history and origins of 68
  influence and reputation of 23
  initial acknowledgement of problems 75–226
  lack of testing by 24, 37, 48, 53, 58–59, 79, 122
  letter on AL-CU markings 37, 47
  meeting on crisis (1971) 43, 52–55, 153, 162, 219
  NEC recommendations and 86
  pigtailing questionnaire 123–124
  pressure for action by 49, 51, 52, 55–56, 58, 70, 111–112
  Research Bulletin #48 69, 180
  skirting the issue of fire hazard 134
  20/20 show and 92
  wire wrap 161
Underwriter's Electrical Bureau 68
Unsafe at Any Speed (Nader) 218
U.S. Department of Agriculture 68
U.S. Deptartment of Health, Education, and Welfare 58
U.S. Rubber 32, 34, 69, 74
Utah
  ban in 55
  pigtailing requirements 80

**V**

Van Deerlin, Lionel 96
VandeWiele, A.G. 78
Vanover, Marsha 10
Vanover, Roberta 10
Virginia, bans in 84

**W**

Waite, Schneider, Bayless and Chesley 190
Walker meter terminal, testing 34
Washington
  ban in 51
  Policy Letter No. 8 on pigtailing 79–80
Watt, J.H. 220
Wedekind, O.G.
  Arizona ban testimony 85, 86
  testing by 81–82, 86, 108, 161, 163
Weitzman, M.J. 47, 114, 122, 165, 167
Westinghouse 213
Whitaker, H.B. (Baron)
  ad hoc committee and 109, 112–114, 118–124, 129
  on pigtailing 120
  responsibility, denial of UL's 116
  responsibility for UL actions 49, 71–74
  Wedekind tests and 83
Wiley, R.O. 121
Wilhoit, Henry R. 194, 195, 196–198
Williamson, J.B.P. 106
Wilson Commission 183–185
Wilson, John Tuzo 183
Wooley, G. 97–98
World Trade Center 92
World War II 29, 31

WRC-TV 97–98
Wright-Malta Corp. 144–146, 182, 183, 223

## Y
Young, Bruce 88
Young fire 88

## Z
zinc plated terminal screws 48

www.ingramcontent.com/pod-product-compliance
Lightning Source LLC
Chambersburg PA
CBHW051353290426
44108CB00015B/1999